Industrial and Applied Mathematics

The **Industrial and Applied Mathematics** series publishes high-quality research-level monographs, lecture notes, textbooks, contributed volumes, focusing on areas where mathematics is used in a fundamental way, such as industrial mathematics, bio-mathematics, financial mathematics, applied statistics, operations research and computer science.

K. Balachandran

An Introduction to Fractional Differential Equations

 Springer

K. Balachandran
Department of Mathematics
Bharathiar University
Coimbatore, Tamil Nadu, India

ISSN 2364-6837   ISSN 2364-6845 (electronic)
Industrial and Applied Mathematics
ISBN 978-981-99-6079-8   ISBN 978-981-99-6080-4 (eBook)
https://doi.org/10.1007/978-981-99-6080-4

Mathematics Subject Classification: 26A33, 33E22, 34A08, 34A12, 35A01, 35R11, 93B05, 93B07, 93D20

This Springer imprint is published by the registered company Springer Nature Singapore Pte Ltd.
The registered company address is: 152 Beach Road, #21-01/04 Gateway East, Singapore 189721,
Singapore

Paper in this product is recyclable.

Fractional Calculus is the calculus of the 21st century.

Fractional Differential Equations are alternative models of nonlinear differential equations.

Preface

Fractional calculus and fractional differential equations are widely used by many researchers in different areas of physical sciences, life sciences, engineering and technology. So there is a need to have an introductory level book on the elementary concepts of fractional calculus and fractional differential equations along with applications for the beginners of mathematics students. There are many books, monographs, lecture notes, and review articles available on the topics but many of them are having less examples and cumbersome notations which discourage the beginners. In order to avoid the difficulty and fulfill the requirement, I have attempted to write an introductory level book on these subjects with an interesting application area. Several new definitions of fractional integrals and fractional derivatives are listed from the literature so that the students can understand the importance and developments of this new area. Further, in each chapter, a set of exercises is included for the benefit of the learners. Although references are made with some of the available books on this subject, the presentation of the material is entirely new and easily comprehensible to the students. In the process some contents of the subjects are refined and examples are added at appropriate levels so that the book will motivate the students to understand the concepts easily and clearly. I hope the students will learn the fundamental concepts and comprehend the contents thoroughly from this book.

I would like to thank my former research students, all of them are now teachers in various institutions in India, and colleagues for reading the manuscript and suggesting improvements of the text. Finally, I thank Mr. Shamim Ahmad, Executive Editor, Springer-India for his continuous encouragement to complete the project in spite of inordinate delay on my part.

Coimbatore, India   K. Balachandran

Contents

Chapter 1
Introduction

Abstract In this chapter, we provide a brief history of fractional calculus with tautochrone problem. Special functions like gamma function, Mittag-Leffler function, Wright function, and Mittag-Leffler matrix function are introduced. The Laplace transform, inverse Laplace transform, and some basic fixed point theorems and function spaces are given. Finally, a set of exercises is provided.

Keywords Tautochrone problem · Gamma function · Mittag-Leffler function · Wright function · Laplace transform · Fixed point theorems

1.1 Brief History

The subject of fractional calculus deals with the investigations of derivatives and integrals of any arbitrary real or complex order which unify and extend the notions of integer order derivative and n-fold integral. It can be considered as a branch of mathematical analysis which deals with integrodifferential operators and equations where the integrals are of convolution type and exhibit (weakly singular) kernels of power-law type. It is strictly related to the theory of pseudo-differential operators. It was found that various especially interdisciplinary applications can be elegantly modeled with the help of the fractional derivatives. In a letter dated September 30, 1695, L' Hospital wrote to Leibniz asking him a particular notation that he had used in his publication for the n-th derivative of a function

$$\frac{d^n f(x)}{dx^n}.$$

What would the result be if $n = \frac{1}{2}$?. Leibniz responded that it would be "an apparent paradox from which one day useful consequences will be drawn". In these words, fractional calculus was born. Studies over the intervening period have proved at least half right. It is clear that in the twentieth century especially numerous applications have been found. However, these applications and mathematical background surrounding fractional calculus are far from paradoxical. While the physical meaning is difficult to grasp, the definitions are no more rigorous than their integer order counterpart.

S. F Lacroix was the first to start the fractional calculus around 1819 by defining the n-th derivative of the power function $y = x^m$ where m is a positive integer as

$$\frac{d^n y}{dx^n} = \frac{m!}{(m-n)!} x^{m-n}, \quad m \geq n$$

and using Legendre's symbol Γ, for the generalized factorial, he wrote

$$\frac{d^n y}{dx^n} = \frac{\Gamma(m+1)}{\Gamma(m-n+1)} x^{m-n},$$

where $m \geq n$. He then let n be any real number to arrive at the definition of a fractional derivative of a power function. By letting $m = 1$ and $n = \frac{1}{2}$, he obtained

$$\frac{d^{\frac{1}{2}} y}{dx^{\frac{1}{2}}} = \frac{2\sqrt{x}}{\sqrt{\pi}}.$$

Three years later, Fourier defined the derivative of arbitrary order in terms of its Fourier transform. Niels Henrik Abel worked with fractional operations on the tautochrone problem in 1823. He was the first to solve an integral equation by means of the fractional calculus. Perhaps, even more importantly, the derivation below will provide an example of how the Riemann–Liouville fractional integral arises in the formulation of physical problems.

Tautochrone Problem: Assume that a thin wire C is placed in the first quadrant of a vertical plane and that a frictionless bead slides along the wire under the action of gravity (see Fig. 1.1). Let the initial velocity of the bead be zero. The tautochrone problem is to determine the shape of the curve such that the time of descent of a frictionless point mass sliding down the curve under the action of gravity is independent of the starting point. This problem should not be confused with the brachistochrone problem in the calculus of variations as that problem is to find the shape of the curve such that the time of descent of the bead from one point to the other would be a minimum.

Now, moving on to the formulation of the problem, let s be the arc length measured along the curve C from O to an arbitrary point Q on C and let α be the angle of inclination. Then $-g \cos \alpha$ is the acceleration $d^2 s / dt^2$ of the bead, where g is the gravitational constant and

$$\frac{d\eta}{ds} = \cos \alpha.$$

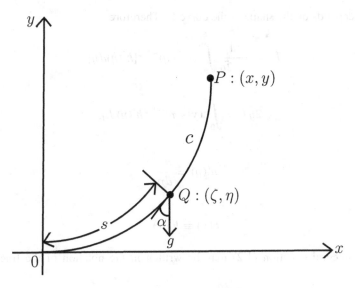

Fig. 1.1 Tautochrone curve

Hence we have the differential equation

$$\frac{d^2s}{dt^2} = -g\frac{d\eta}{ds}.$$

With the aid of the integrating factor ds/dt, we see immediately that

$$\left(\frac{ds}{dt}\right)^2 = -2g\eta + k, \qquad (1.1)$$

where k is a constant of integration. Since the bead started from rest, ds/dt is zero when $\eta = y$ and thus $k = 2gy$. Therefore (1.1) can be written as

$$\frac{ds}{dt} = -\sqrt{2g(y - \eta)}.$$

The negative square root is chosen since as t increases, s decreases.

Thus the time of descent T from P to O is

$$T = -\frac{1}{\sqrt{2g}} \int_P^O \frac{1}{\sqrt{y - \eta}} ds.$$

Now the arc length s is a function of η, say,

$$s = h(\eta),$$

where h depends on the shape of the curve C. Therefore

$$T = -\frac{1}{\sqrt{2g}} \int_y^0 (y - \eta)^{-1/2} [h'(\eta)d\eta]$$

or

$$\sqrt{2g}T = \int_0^y (y - \eta)^{-1/2} h'(\eta)d\eta, \tag{1.2}$$

where

$$h'(\eta) = \frac{ds}{d\eta}. \tag{1.3}$$

If we let

$$f(y) \equiv h'(y), \tag{1.4}$$

then the integral equation (1.2) may be written in the notation of the fractional calculus as

$$\frac{\sqrt{2g}}{\Gamma(\frac{1}{2})}T = D^{-1/2}f(y). \tag{1.5}$$

But the right-hand side of (1.5) is the Riemann–Liouville fractional integral of f of order $\frac{1}{2}$. This is our desired formulation. It remains to solve (1.5) and then to find the equation of C.

Abel applied the fractional operator $D^{1/2}$ to both sides of (1.5) and thereof to have

$$D^{1/2}\sqrt{\frac{2g}{\pi}}T = f(y). \tag{1.6}$$

Now we know from [1] that this is legitimate if f and T are of class C. But a constant is certainly of class C and since

$$D^{1/2}T = \frac{T}{\sqrt{\pi y}},$$

we see that f also is of class C. Thus (1.6) becomes

$$f(y) = \frac{\sqrt{2g}}{\pi}Ty^{-1/2}, \tag{1.7}$$

which is the solution of (1.5) [or (1.2)].

We also could have solved (1.5) by the Laplace transform technique since (1.2) is a convolution integral.

Now to solve the second part of the problem, that is, to find the equation of C, we begin by using (1.4) and (1.3) to write

$$f(y) = h'(y) = \frac{ds}{dy} = \sqrt{1 + \left(\frac{dx}{dy}\right)^2}.$$

Thus

$$\frac{dx}{dy} = \sqrt{f^2(y) - 1}$$

or

$$x = \int_0^y \sqrt{\left(\frac{2gT^2}{\pi^2\eta} - 1\right)} d\eta + c. \qquad (1.8)$$

But $c = 0$ since at the origin $x = 0 = y$.

If we let

$$a = \frac{gT^2}{\pi^2},$$

then the change of variable of integration

$$\eta = 2a \sin^2 \zeta$$

reduces (1.8) to

$$x = 4a \int_0^\beta \cos^2 \zeta d\zeta,$$

where

$$\beta = \arcsin \sqrt{\frac{y}{2a}}.$$

These last two equations then imply that

$$x = 2a(\beta + \frac{1}{2} \sin 2\beta),$$
$$y = 2a \sin^2 \beta,$$

and if we make the trivial change of variable $\theta = 2\beta$, the parametric equations of C become

$$\left. \begin{array}{l} x = a(\theta + \sin \theta) \\ y = a(1 - \cos \theta) \end{array} \right\} \qquad (1.9)$$

with $a = \frac{gT^2}{\pi^2}$. The solution of the problem is now complete.

We see from (1.9) that C is a cycloid. One may also formulate the more general problem of determining C such that the time of descent T, instead of being a constant, is a specified function of η, say $\psi(\eta)$. Then (1.5) becomes

$$\sqrt{\frac{2g}{\pi}}\psi(y) = D^{-1/2}f(y),$$

and under suitable conditions on ψ, the solution of the fractional integral equation above is

$$f(y) = \sqrt{\frac{2g}{\pi}}D^{1/2}\psi(y).$$

Although this problem may seem to be a simple exercise in elementary mechanics and differential equations, it turned out to be of greater mathematical significance. Even though the tautochrone problem was attacked and solved by mathematicians long before Abel, it was he who first solved it by means of the fractional calculus.

1.2   Special Functions

(i) Gamma Function:

The gamma function is defined by the Euler integral of the second kind as

$$\Gamma(z) = \int_0^\infty t^{z-1}e^{-t}dt, \quad \mathbb{R}(z) > 0. \tag{1.10}$$

However we have to make sure that this definition makes sense, that is, the improper integral converges for all complex $z \in \mathbb{C}, \mathbb{R}(z) > 0$, real part of z. For this function, the reduction formula

$$\Gamma(z+1) = z\Gamma(z), \mathbb{R}(z) > 0 \tag{1.11}$$

holds; it is obtained from (1.10) by integration by parts. Using this relation, the Euler gamma function is extended to the half-plane $\mathbb{R}(z) \leq 0$ by

$$\Gamma(z) = \frac{\Gamma(z+n)}{(z)_n}, \quad \mathbb{R}(z) > -n, n \in \mathbb{N}, z \notin \mathbb{Z}_0^- := 0, -1, -2\cdots \tag{1.12}$$

Here $(z)_n$ is the Pochhammer symbol, defined for complex $z \in \mathbb{C}$ and non-negative integer $n \in \mathbb{N}_0$ by

$$(z)_0 = 1 \text{ and } (z)_n = z(z+1)\cdots(z+n-1), n \in \mathbb{N}. \tag{1.13}$$

Equations (1.12) and (1.13) yield

$$\Gamma(n+1) = (1)_n = n!, \ n \in \mathbb{N}_0$$

with $0! = 1$. It follows from (1.12) that the gamma function is analytic everywhere in the complex plane $\mathbb{C}$ except at $z = 0, -1, -2, \cdots$, where $\Gamma(z)$ has simple poles and is represented by the asymptotic formula

$$\Gamma(z) = \frac{(-1)^k}{z + k}[1 + O(z + k)], z \to -k, k \in \mathbb{N}_0.$$

We also indicate some other properties of the gamma function such as the functional equation:

$$\Gamma(z)\Gamma(1 - z) = \frac{\pi}{\sin \pi z}, \ z \notin \mathbb{Z}_0; 0 < \mathbb{R}(z) < 1,$$

and

$$\Gamma(\frac{1}{2}) = \sqrt{\pi}.$$

Alternatively we now show that $\Gamma(\frac{1}{2}) = \sqrt{\pi}$. By definition we have

$$\Gamma(\frac{1}{2}) = \int_0^\infty e^{-t} t^{-\frac{1}{2}} dt.$$

If we let $t = y^2$, then $dt = 2y dy$ and we now have

$$\Gamma(\frac{1}{2}) = 2 \int_0^\infty e^{-y^2} dy.$$

Equivalently we can write as

$$\Gamma(\frac{1}{2}) = 2 \int_0^\infty e^{-x^2} dx.$$

If we multiply both, then we get

$$[\Gamma(\frac{1}{2})]^2 = 4 \int_0^\infty \int_0^\infty e^{-(x^2 + y^2)} dx dy.$$

Using polar coordinates $x = r \cos \theta$ and $y = r \sin \theta$, we get

$$[\Gamma(\frac{1}{2})]^2 = 4 \int_0^{\frac{\pi}{2}} \int_0^\infty e^{-r^2} r dr d\theta = \pi.$$

Thus $\Gamma(\frac{1}{2}) = \sqrt{\pi}$.

Some more properties of gamma function are as follows:

- $\Gamma(z) = \frac{\Gamma(z+1)}{z}$, for negative value of z.
- $\Gamma(\frac{-1}{2}) = \frac{\Gamma(\frac{-1}{2}+1)}{\frac{-1}{2}} = -2\Gamma(\frac{1}{2}) = -2\sqrt{\pi}$.
- $\Gamma(\frac{-3}{2}) = \frac{\Gamma(\frac{-3}{2}+1)}{\frac{-3}{2}} = \frac{4}{3}\sqrt{\pi}$.
- $\Gamma(x) = \lim_{n\to\infty} \frac{n!}{x(x+1)\cdots(x+n)} n^x$.

(ii) Mittag-Leffler Function:

In 1903, the Swedish mathematician Gosta Mittag-Leffler introduced the function $E_\alpha(z)$ defined by

$$E_\alpha(z) = \sum_{n=0}^{\infty} \frac{z^n}{\Gamma(\alpha n + 1)}, \tag{1.14}$$

where z is a complex variable and $\alpha \geq 0$ (for $\alpha = 0$ the radius of convergence of the sum (1.14) is finite and one has by definition $E_0(z) = 1/(1-z)$). The Mittag-Leffler function is a direct generalization of the exponential function because of substitution of $n! = \Gamma(n+1)$ with $(\alpha n)! = \Gamma(\alpha n + 1)$. For $0 < \alpha < 1$, it interpolates between the pure exponential and a hypergeometric function $\frac{1}{1-z}$. Its importance is realized in the problems of physics, chemistry, biology, engineering, and applied sciences because Mittag-Leffler function naturally occurs as the solution of fractional differential equations or fractional integral equations.

The generalization of $E_\alpha(z)$ was studied by Wiman in 1905 and he defined the function

$$E_{\alpha,\beta}(z) = \sum_{n=0}^{\infty} \frac{z^n}{\Gamma(\alpha n + \beta)}, \tag{1.15}$$

where $z, \alpha, \beta \in \mathbb{C}$ and $\Re(\alpha) > 0$, $\Re(\beta) > 0$, which is called as Wiman's function or generalized Mittag-Leffler function as $E_{\alpha,1}(z) = E_\alpha(z)$. By definition

$$E_{\alpha,\beta}(z) = \sum_{n=0}^{\infty} \frac{z^n}{\Gamma(\alpha n + \beta)}$$

$$= \sum_{n=-1}^{\infty} \frac{z^{n+1}}{\Gamma(\alpha n + \alpha + \beta)}$$

$$= \frac{1}{\Gamma(\beta)} + z\sum_{n=0}^{\infty} \frac{z^n}{\Gamma(\alpha n + \alpha + \beta)}$$

$$= \frac{1}{\Gamma(\beta)} + zE_{\alpha,\alpha+\beta}(z).$$

Note that $E_\alpha(z_1 + z_2) \neq E_\alpha(z_1)E_\alpha(z_2)$ and $E_{\alpha,\beta}(z_1 + z_2) \neq E_{\alpha,\beta}(z_1)E_{\alpha,\beta}(z_2)$ and the equality occurs only when $\alpha = \beta = 1$. From the definition, we have the following identities:

$$E_1(z) = \sum_{k=0}^{\infty} \frac{z^k}{\Gamma(k+1)} = \sum_{k=0}^{\infty} \frac{z^k}{k!} = e^z,$$

$$E_{1,2}(z) = \sum_{k=0}^{\infty} \frac{z^k}{\Gamma(k+2)} = \frac{1}{z}\sum_{k=0}^{\infty} \frac{z^{k+1}}{(k+1)!} = \frac{e^z - 1}{z},$$

$$E_{1,3}(z) = \sum_{k=0}^{\infty} \frac{z^k}{\Gamma(k+3)} = \frac{1}{z^2}\sum_{k=0}^{\infty} \frac{z^{k+2}}{(k+2)!} = \frac{e^z - 1 - z}{z^2}.$$

From two-parameter Mittag-Leffler function, we can obtain the trigonometric and hyperbolic functions as follows:

$$E_{2,1}(z^2) = \sum_{k=0}^{\infty} \frac{z^{2k}}{\Gamma(2k+1)} = \cosh(z),$$

$$E_{2,2}(z^2) = \frac{1}{z}\sum_{k=0}^{\infty} \frac{z^{2k+1}}{(2k+1)!} = \frac{\sinh(z)}{z}.$$

The Mittag-Leffler function (1.15) can be expressed in the integral representation for $z \in \mathbb{C}, \alpha > 0$, as

$$E_\alpha(z) = \frac{1}{2\pi i}\int_C \frac{t^{\alpha-1}e^t}{t^\alpha - z}dt,$$

$$E_{\alpha,\beta}(z) = \frac{1}{2\pi i}\int_C \frac{t^{\alpha-\beta}}{t^\alpha - z}e^t dt,$$

where the path of integration C starts and ends at $-\infty$ and encircles the circular disk $|t| \leq |z|^\alpha$. Also we have the following identities:

$$E_{\alpha,\beta}(z) + E_{\alpha,\beta}(-z) = 2E_{2\alpha,\beta}(z^2)$$
$$E_{\alpha,\beta}(z) - E_{\alpha,\beta}(-z) = 2z E_{2\alpha,\alpha+\beta}(z^2).$$

Next we show that $\frac{d}{dz}E_\beta(z) = \frac{1}{\beta}E_{\beta,\beta}(z)$ for $0 < \beta < \infty$. For, the Mittag-Leffler function is analytic and thus absolutely convergent. Hence we may interchange the sum and derivative and calculate the derivative of $E_\beta(z)$ as

$$\frac{d}{dz}E_\beta(z) = \frac{d}{dz}\sum_{n=0}^{\infty}\frac{z^n}{\Gamma(\beta n + 1)}$$

$$= \sum_{n=1}^{\infty}\frac{nz^{n-1}}{\Gamma(\beta n + 1)}$$

$$= \sum_{n=0}^{\infty}\frac{(n+1)z^n}{\Gamma(\beta(n+1)+1)}$$

$$= \sum_{n=0}^{\infty}\frac{(n+1)z^n}{\beta(n+1)\Gamma(\beta(n+1))}$$

$$= \frac{1}{\beta}\sum_{n=0}^{\infty}\frac{z^n}{\Gamma(\beta n + \beta)}$$

$$= \frac{E_{\beta,\beta}(z)}{\beta}.$$

Note that [2]

$$\int_0^\infty e^{-t}t^{\beta-1}E_{\alpha,\beta}(zt^\alpha)dt = \frac{1}{1-z}, \quad |z| < 1$$

and

$$\frac{d^k}{dz^k}E_k(z^k) = E_k(z^k).$$

(iii) Wright function:

Next we define a new type of function called Wright function as

$$W_{\alpha,\beta}(z) = \sum_{k=0}^{\infty}\frac{z^k}{k!\Gamma(\alpha k + \beta)}, \quad \alpha > -1, \quad \beta \in \mathbb{C}.$$

Observe that

$$W_{0,\beta}(z) = \frac{e^z}{\Gamma(\beta)}$$

and

$$\frac{d}{dz}W_{\alpha,\beta}(z) = W_{\alpha,\alpha+\beta}(z).$$

Consider the function of Wright type

$$W_\alpha(z) = \sum_{n=0}^{\infty} \frac{(-z)^n}{n!\Gamma(-\alpha n + 1 - \alpha)}$$

$$= \frac{1}{\pi} \sum_{n=1}^{\infty} \frac{(-z)^{n-1}}{(n-1)!} \Gamma(n\alpha) \sin(n\pi\alpha), \quad z \in \mathbb{C},$$

with $0 < \alpha < 1$. For $-1 < r < \infty$, $\lambda > 0$, the following results hold:

(i) $\int_0^\infty \frac{\alpha}{t^{\alpha+1}} W_\alpha(\frac{1}{t^\alpha}) e^{-\lambda^\alpha} dt = e^{-\lambda^\alpha}$,

(ii) $\int_0^\infty W_\alpha(t) t^r dt = \frac{\Gamma(1+r)}{\Gamma(1+\alpha r)}$,

(iii) $\int_0^\infty W_\alpha(t) e^{-zt} dt = E_\alpha(-z), \quad z \in \mathbb{C}$,

(iv) $\int_0^\infty \alpha t W_\alpha(t) e^{-zt} dt = E_{\alpha,\alpha}(-z), \quad z \in \mathbb{C}$.

Another generalized form of Mittag-Leffler function is defined by

$$E_{\alpha,\beta}^\gamma(z) = \sum_{k=0}^{\infty} \frac{(\gamma)_k z^k}{k!\Gamma(\alpha k + \beta)}, \tag{1.16}$$

where $(\gamma)_k$ is a Pochhammer symbol. Its integral representation is

$$E_{\alpha,\beta}^\gamma(z) = \frac{1}{2\pi i} \int_C e^s \frac{s^{\alpha\gamma-\beta}}{(s^\alpha - z)^\gamma} ds,$$

where C is any suitable contour in the complex plane encompassing at the left all the singularities of the integrand.

Also we have [2]

$$\frac{d^k}{dx^k} E_{\alpha,\beta}(x) = k! E_{\alpha,\beta+\alpha k}^{k+1}(x)$$

and

$$\frac{d^k}{dx^k} E_{\alpha,\beta}^\gamma(x) = (\gamma)_k E_{\alpha,\beta+\alpha k}^{\gamma+k}(x).$$

(iv) Mittag-Leffler Matrix Function:

If A is an $n \times n$ matrix, then the Mittag-Leffler matrix function $E_\alpha(At^\alpha)$ is defined as

$$E_\alpha(At^\alpha) = \sum_{k=0}^{\infty} \frac{A^k t^{\alpha k}}{\Gamma(\alpha k + 1)}, \quad \alpha > 0, t \in \mathbb{R}. \tag{1.17}$$

The two-parameter Mittag-Leffler matrix function is defined as

$$E_{\alpha,\beta}(At^\alpha) = \sum_{k=0}^{\infty} \frac{A^k t^{\alpha k}}{\Gamma(\alpha k + \beta)}, \quad \alpha, \beta > 0, t \in \mathbb{R}. \tag{1.18}$$

Now we calculate $E_\alpha(At^\alpha)$, when $A = \begin{pmatrix} a & -b \\ b & a \end{pmatrix}$. By definition we have

$$E_\alpha(At^\alpha) = \sum_{k=0}^{\infty} \frac{A^k t^{\alpha k}}{\Gamma(\alpha k + 1)}.$$

However it is easy to see that

$$A^k = \begin{pmatrix} a & -b \\ b & a \end{pmatrix}^k = \begin{pmatrix} Re(\lambda^k) & -Im(\lambda^k) \\ Im(\lambda^k) & Re(\lambda^k) \end{pmatrix}.$$

Therefore

$$E_\alpha(At^\alpha) = \sum_{k=0}^{\infty} \frac{t^{\alpha k}}{\Gamma(\alpha k + 1)} \begin{pmatrix} Re(\lambda^k) & -Im(\lambda^k) \\ Im(\lambda^k) & Re(\lambda^k) \end{pmatrix}.$$

On the other hand, if we write $\lambda = re^{i\theta}$, where $r = \sqrt{a^2 + b^2}$ and $\theta = arctg(\frac{a}{b})$, then $\lambda^k = r^k e^{ik\theta} = r^k(\cos(k\theta) + i\sin(k\theta))$ and therefore

$$E_\alpha(At^\alpha) = \sum_{k=0}^{\infty} \frac{r^k t^{\alpha k}}{\Gamma(\alpha k + 1)} \begin{pmatrix} \cos(k\theta) & -\sin(k\theta) \\ \sin(k\theta) & \cos(k\theta) \end{pmatrix}.$$

Next we state the properties of Mittag-Leffler matrix function. If A, B are $n \times n$ matrices, then we have the following properties[3]:

  (i) $AE_{\alpha,\beta}(A) = E_{\alpha,\beta}(A)A$.
 (ii) $E_{\alpha,\beta}(A^*) = (E_{\alpha,\beta}(A))^*$;(* denotes matrix transpose).
(iii) $E_{\alpha,\beta}(XAX^{-1}) = XE_{\alpha,\beta}(A)X^{-1}$ for any nonsingular matrix X.
 (iv) The eigenvalues of $E_{\alpha,\beta}(A)$ are $E_{\alpha,\beta}(\lambda_i)$ where λ_i are the eigenvalues of A.
  (v) If B commutes with A, then B commutes with $E_{\alpha,\beta}(A)$.
 (vi) $E_{\alpha,\beta}(AB)A = AE_{\alpha,\beta}(BA)$.
(vii) $E_{\alpha,\beta}(0) = \frac{1}{\Gamma(\beta)}I$ where I and 0 are the identity and zero matrices of dimension n.
(viii) $A^m E_{\alpha,\beta+m\alpha}(A) = E_{\alpha,\beta}(A) - \sum_{k=0}^{m-1} \frac{A^k}{\Gamma(\alpha k + \beta)}$.
 (ix) In general $E_\alpha(A + B) \neq E_\alpha(A)E_\alpha(B)$, but if A and B are nilpotent with index 2 and $AB = BA = 0$, then $E_\alpha(A + B) = E_\alpha(A)E_\alpha(B)$.

(x) $E_{\alpha,\beta}(A) = \dfrac{1}{2\pi i}\displaystyle\int_C E_{\alpha,\beta}(z)(zI - A)^{-1}dz$, where C is a closed contour enclos-
ing the spectrum of A.

1.3 Laplace Transform

Suppose that f is a real or complex valued function of the (time) variable $t > 0$ and s is a real or complex parameter. We define the Laplace transform of f as

$$F(s) = \mathcal{L}(f(t)) = \int_0^\infty e^{-st} f(t)dt = \lim_{T\to\infty}\int_0^T e^{-st} f(t)dt$$

whenever the limit exists. When it does, the integral is said to converge. If the limit does not exist, the integral is said to diverge and there is no Laplace transform defined for f. The notation $\mathcal{L}(f)$ will also be used to denote the Laplace transform of f and the integral is the ordinary Riemann integral. The parameter s belongs to some domain on the real line or in the complex plane. Let $f(t) = e^{at}$, a real. This function is continuous on $[0, \infty)$ and of exponential order a. Then

$$\mathcal{L}(e^{at}) = \int_0^\infty e^{-st} e^{at} dt,$$

$$= \int_0^\infty e^{-(s-a)t} dt,$$

$$= \left(\frac{e^{-(s-a)t}}{-(s-a)}\right)_0^\infty = \frac{1}{(s-a)}, \quad \mathbb{R}(s) > a.$$

The same calculation holds for complex a and $\mathbb{R}(s) > \mathbb{R}(a)$.
Let $f(t) = t$ which is continuous and of exponential order, then

$$\mathcal{L}(t) = \int_0^\infty t e^{-st} dt = \frac{1}{s^2}.$$

Performing integration by parts twice as above, we find that

$$\mathcal{L}(t^2) = \int_0^\infty e^{-st} t^2 dt = \frac{2}{s^3}, \quad \mathbb{R}(s) > 0.$$

By induction, one can show that in general

$$\mathcal{L}(t^n) = \frac{n!}{s^{n+1}}, \quad \mathbb{R}(s) > 0,$$

for $n = 1, 2, 3 \cdots$. Indeed this formula holds for $n = 0$, since $0! = 1$.

The Laplace transform of one parameter Mittag-Leffler function is defined as

$$\mathcal{L}[E_\alpha(\pm\lambda t^\alpha)](s) = \int_0^\infty e^{-st} \sum_{k=0}^\infty \frac{\pm\lambda^k t^{\alpha k}}{\Gamma(\alpha k + 1)} dt,$$

$$= \sum_{k=0}^\infty \frac{\pm\lambda^k}{\Gamma(\alpha k + 1)} \int_0^\infty e^{-st} t^{\alpha k} dt,$$

$$= \sum_{k=0}^\infty \frac{\pm\lambda^k}{\Gamma(\alpha k + 1)} \cdot \frac{1}{s} \cdot \frac{1}{s^\alpha k} \Gamma(\alpha k + 1),$$

$$= \frac{1}{s} \sum_{k=0}^\infty \frac{\pm\lambda^k}{s^\alpha k},$$

$$= \frac{1}{s} [1 + \frac{\pm\lambda}{s^\alpha} + \frac{\pm\lambda^2}{s^{2\alpha}} + \cdots],$$

$$= \frac{1}{s} [(1 - \frac{\pm\lambda}{s^\alpha})^{-1}],$$

$$= \frac{s^{\alpha-1}}{s^\alpha \pm \lambda}.$$

The Laplace transform of two-parameter Mittag-Leffler function is defined as

$$\mathcal{L}[t^{\beta-1} E_{\alpha,\beta}](\pm\lambda t^\alpha)(s) = \frac{s^{\alpha-\beta}}{s^\alpha \pm \lambda},$$

where $\mathbb{R}(\alpha) > 0$, $\mathbb{R}(\beta) > 0$ and the Laplace transform of generalized Mittag-Leffler function is defined as

$$\mathcal{L}[t^{\beta-1} E_{\alpha,\beta}^{(\gamma)}(\pm\lambda t^\alpha)](s) = \frac{s^{\alpha\gamma-\beta}}{(s^\alpha \pm \lambda)^\gamma}. \mathbb{R}(s) > 0, \mathbb{R}(\beta) > 0, |\lambda s^{-\alpha}| < 1.$$

The Laplace transform of the convolution

$$f(t) * g(t) = \int_0^t f(t - \tau) g(\tau) d\tau,$$

$$= \int_0^t f(\tau) g(t - \tau) d\tau,$$

of the two functions $f(t)$ and $g(t)$, which are equal to zero for $t < 0$, is equal to the product of the Laplace transforms of those functions

$$\mathcal{L}[f(t) * g(t)](s) = F(s) G(s),$$

under the assumption that both $F(s)$ and $G(s)$ exist.

1.4 Inverse Laplace Transform

If $\mathcal{L}\{f(t)\} = F(s)$, then the inverse Laplace transform of $F(s)$ is defined as

$$\mathcal{L}^{-1}\{F(s)\} = f(t).$$

The inverse transform $\mathcal{L}^{-1}$ is a linear operator:

$$\mathcal{L}^{-1}\{F(s) + G(s)\} = \mathcal{L}^{-1}\{F(s)\} + \mathcal{L}^{-1}\{G(s)\},$$
$$\mathcal{L}^{-1}\{cF(s)\} = c\mathcal{L}^{-1}\{F(s)\},$$

for any constant c.
As an example, the inverse Laplace transform of

$$F(s) = \frac{1}{s^3} + \frac{6}{s^2 + 4}$$

is

$$f(t) = \mathcal{L}^{-1}\{F(s)\}$$
$$= \frac{1}{2}\mathcal{L}^{-1}\left\{\frac{2}{s^3}\right\} + 3\mathcal{L}^{-1}\left\{\frac{2}{s^2 + 4}\right\}$$
$$= \frac{t^2}{2} + 3\sin 2t.$$

There is usually more than one way to invert the Laplace transform. For example, also let $F(s) = (s^2 + 4s)^{-1}$. We can compute the inverse transform of this function by completing the square:

$$f(t) = \mathcal{L}^{-1}\left\{\frac{1}{s^2 + 4s}\right\}$$
$$= \mathcal{L}^{-1}\left\{\frac{1}{(s + 2)^2 - 4}\right\}$$
$$= \frac{1}{2}\mathcal{L}^{-1}\left\{\frac{2}{(s + 2)^2 - 4}\right\}$$
$$= \frac{1}{2}e^{-2t}\sinh 2t,$$

and use the partial fraction decomposition of $F(s)$ as

$$F(s) = \frac{1}{s(s + 4)} = \frac{1}{4s} - \frac{1}{4(s + 4)}.$$

Therefore

$$f(t) = \mathcal{L}^{-1}\{F(s)\}$$

$$= \mathcal{L}^{-1}\left\{\frac{1}{4s}\right\} - \mathcal{L}^{-1}\left\{\frac{1}{4(s+4)}\right\}$$

$$= \frac{1}{4} - \frac{1}{4}e^{-4t}$$

$$= \frac{1}{2}e^{-2t}\sinh 2t.$$

Next we find the inverse Laplace transform $q(t)$ of $Q(s) = \frac{3s}{(s^2+1)^2}$. Note that

$$Q(s) = -\frac{3}{2}\frac{d}{ds}\frac{1}{s^2+1}.$$

Hence

$$q(t) = \mathcal{L}^{-1}\{Q(s)\}$$

$$= -\frac{3}{2}\mathcal{L}^{-1}\left\{\frac{d}{ds}\frac{1}{s^2+1}\right\}$$

$$= \frac{3}{2}t\sin t.$$

If the Laplace transforms of $f(t)$ and $g(t)$ are $F(s)$ and G(s), respectively, then

$$\mathcal{L}\{(f*g)(t)\} = F(s)G(s),$$

and so

$$\mathcal{L}^{-1}\{F(s)G(s)\} = (f*g)(t).$$

Suppose we want to find the inverse transform $x(t)$ of $X(s)$. Write $X(s)$ as a product $F(s)G(s)$ where $f(t)$ and $g(t)$ are known, then by the above result, $x(t) = (f*g)(t)$. For the inverse transform $q(t)$ of $Q(s) = \frac{3s}{(s^2+1)^2}$. We write $Q(s) = F(s)G(s)$, where $F(s) = \frac{3}{s^2+1}$ and $G(s) = \frac{s}{s^2+1}$. But the inverses of $F(s)$ and G(s) are $f(t) = 3\sin t$ and $g(t) = \cos t$. Therefore

$$q(t) = \mathcal{L}^{-1}\{Q(s)\}$$

$$= \mathcal{L}^{-1}\{F(s)G(s)\}$$

$$= (f*g)(t)$$

$$= 3\int_0^t \sin(t-v)\cos v\,dv.$$

By using the trigonometric identity $2 \sin A \cos B = \sin(A + B) + \sin(A - B)$, we have

$$q(t) = \frac{3}{2} \int_0^t \sin t \, dv + \int_0^t \sin(t - 2v) dv$$

$$= \frac{3}{2} t \sin t.$$

For finding the inverse Laplace transform $x(t)$ of the function $X(s)$, we can use the convolution theorem and write as

$$X(s) = \frac{1}{s} \frac{1}{s^2 + 4}.$$

Since $\mathcal{L}^{-1}\{\frac{1}{s}\} = 1$ and $\mathcal{L}^{-1}\{\frac{1}{s^2+4}\} = \frac{1}{2} \sin 2t$, we have

$$x(t) = \frac{1}{2} \int_0^t \sin 2v \, dv$$

$$= \frac{1}{4}(1 - \cos 2t).$$

1.5 Fixed Point Theorems

The fixed point technique is one of the powerful methods mainly applied to prove the existence and uniqueness of solutions of differential equations. It is a useful method in different branches of mathematics.

Let X be a Banach space and T be an operator such that $T : X \rightarrow X$. We say that $x \in X$ is a fixed point of T if $Tx = x$. A mapping $T : X \rightarrow X$ is said to be a contraction if there exists a real number α, $0 \leq \alpha < 1$, such that $\|Tx - Ty\| \leq \alpha\|x - y\|$ for all $x, y \in X$. Note that $\|.\|$ indicates a norm in X.

The following theorem, known as Banach's contraction mapping principle, is an important source of proving existence and uniqueness theorems in different branches of analysis.

Theorem 1.5.1 (Banach fixed point theorem) *If X is a Banach space and $T : X \rightarrow X$ is a contraction mapping, then T has a unique fixed point.*

A generalization of the above theorem is as follows.

Theorem 1.5.2 *Let X be a Banach space and let $T : X \rightarrow X$ be such that $T^n : X \rightarrow X$ is a contraction for some n. Then T has a unique fixed point.*

In 1911, Brouwer proved the following fixed point theorem in a finite-dimensional space:

Theorem 1.5.3 (Brouwer fixed point theorem) *Let $D \subset R^n$ be a nonempty compact convex subset and $f : D \to D$ be continuous. Then f has a fixed point.*

Brouwer's theorem fails if the dimension of the space X is infinite, since in infinite-dimensional Banach spaces, bounded sets need not be relatively compact. In 1930, Schauder extended the domain of validity of the Brouwer theorem by proving the following fixed point theorem. The Schauder fixed point theorem will be helpful to assert the existence of solutions of some initial value problems in differential equations. In this theorem, the fixed point lies in a space of functions and this point may be a function that solves a nonlinear integral equation or a partial differential equation.

Now let M be a subset of a Banach space X and A be an operator, generally nonlinear, defined on M and mapping M into itself. The operator A is called *compact* on the set M if it carries every bounded subset of M into a compact set. If, in addition, A is continuous on M (that is, it maps bounded sets into bounded sets), then it is said to be *completely continuous* on M (in case of A being linear, both the definitions coincide).

Theorem 1.5.4 (Schauder Theorem) *Let X be a real Banach space, $M \subset X$ a nonempty closed bounded convex subset and $F : M \to M$ be compact. Then F has a fixed point.*

The following theorems are alternate forms of Schauder's theorem.

Theorem 1.5.5 (Leray–Schauder Theorem) *Every completely continuous operator which maps a closed bounded convex subset of a Banach space into itself has at least one fixed point.*

Theorem 1.5.6 *Let T be a compact continuous mapping of a normed linear space X into X. Then T has a fixed point.*

Theorem 1.5.7 (Schaefer Theorem) *Let E be a normed linear space. Let $F : E \to E$ be a completely continuous operator and let*

$$\zeta(F) = \{x \in E : x = \lambda F x \quad \text{for some} \quad 0 < \lambda < 1\}.$$

Then either $\zeta(F)$ is unbounded or F has a fixed point.

For more about fixed point theorems, one can refer the book by Smart [4].

1.6 Function Spaces

The Space $C(J, \mathbb{R}^n)$

Let $J = [a, b]$ $(-\infty < a < b < \infty)$ be a finite interval on the real line $\mathbb{R}$. Let $C(J, \mathbb{R}^n)$ denote the Banach space of continuous functions $x(t)$ with values in $\mathbb{R}^n$ for $t \in J$ with the norm

$$||x|| \doteq ||x||_C = \sup\{|x(t)| : t \in J\},$$

where $|x(t)|$ is the usual Euclidean norm in $\mathbb{R}^n$.

Let $C^1(J, \mathbb{R}^n)$ denote the Banach space of continuously differentiable functions $x(t)$ with values in $\mathbb{R}^n$ for $t \in J$ with the norm

$$||x||_{C^1} = ||x||_C + ||x'||_C.$$

The functions of a set K are said to be **uniformly bounded** if there is a constant M such that $||f(t)|| \leq M$ for all $f \in K$ and all $t \in [a, b]$. They are called **equicontinuous** if for every $\epsilon > 0$ there is a $\delta > 0$ depending only on ϵ such that $|t_1 - t_2| < \delta$ for every $t_1, t_2 \in [a, b]$ and $||f(t_1) - f(t_2)|| < \epsilon$ for all $f \in K$.

Theorem 1.6.1 (Arzela-Ascoli) *A set $K \subset C(J : \mathbb{R}^n)$ is compact if and only if it is uniformly bounded and equicontinuous.*

The Space $L_p(J : \mathbb{R})$

Let $L_p(J : \mathbb{R})(1 \leq p < \infty)$ denote the space of all p-integrable measurable functions with norm $||f||_p = (\int_a^b |f(x)|^p dx)^{\frac{1}{p}}$ and $L_\infty(J : \mathbb{R})$ be the space of all essentially bounded measurable functions with norm $||f||_\infty = ess \sup_{a \leq x \leq b} |f(x)|$.

Let $L_2(J : \mathbb{R}^n)$ be the space of all measurable n-vector valued functions $f(t)$ defined for $t \in J$ with values in $\mathbb{R}^n$ such that $\int_a^b |f(t)|^2 dt < \infty$. In $L_2(J : \mathbb{R}^n)$, the inner product

$$(f, g)_{L_2(J : \mathbb{R}^n)} = \int_a^b (f(t), g(t))_{\mathbb{R}^n} dt$$

is well defined, $(f(t), g(t))_{\mathbb{R}^n}$ is the usual scalar or dot product in $\mathbb{R}^n$:

$$(f(t), g(t))_{\mathbb{R}^n} = g^*(t) f(t)$$

and each function $f \in L_2(J : \mathbb{R}^n)$ has associated with it a norm

$$||f||_2 = (\int_a^b |f(t)|^2 dt)^{\frac{1}{2}}.$$

Here $*$ denotes the adjoint. The norm of a continuous $n \times m$ matrix valued function $A : J \rightarrow \mathbb{R}^n \times \mathbb{R}^m$ is defined by

$$\|A(t)\| = \max_i \sum_{j=1}^{m} \max |a_{ij}(t)|,$$

where $a_{ij}(t)$ are the elements of $A(t)$.

The Space $AC[a, b]$

A function $f(x)$ is called absolutely continuous on $[a, b]$, if for every $\epsilon > 0$ there exists a $\delta > 0$ such that whenever any finite set of pairwise disjoint intervals $[a_k, b_k] \subset [a, b], k = 1, 2, \cdots, n$ satisfies $\sum_{k=1}^{n}(b_k - a_k) < \delta$ then $\sum_{k=1}^{n} |f(b_k) - f(a_k)| < \epsilon$. The space of these functions is denoted by $AC[a, b]$.

Let $AC[a, b]$ be the space of the absolutely continuous functions f in $[a, b]$ and $AC^n[a, b]$ be the space of the absolutely continuous functions f which have continuous derivative up to order $n - 1$ in $[a, b]$ such that $f^{(n-1)} \in AC[a, b]$.

The space $AC^n[a, b]$ consists of those functions $f(x)$ which can be represented in the form

$$f(x) = I_{a+}^n \phi(x) + \sum_{k=0}^{n-1} c_k (x - a)^k,$$

where $\phi(t) \in L_1(a, b)$, $c_k (k = 0, 1, \cdots, n - 1)$ are arbitrary constants and

$$I_{a+}^n \phi(x) = \frac{1}{(n - 1)!} \int_a^x (x - t)^{n-1} \phi(t) dt.$$

It follows that $\phi(t) = f^n(t)$, $c_k = \frac{f^k(a)}{k!}$ $(k = 0, 1, \cdots, n - 1)$.

Theorem 1.6.2 (Lebesgue Dominated Convergence Theorem) *Let g be integrable over a set E and let $\{f_n\}$ be a sequence of measurable functions such that $|f_n(x)| \leq g(x)$ on E and $f(x) = \lim_{n \to \infty} f_n(x)$ for almost all $x \in E$. Then*

$$\int_E f(x) dx = \lim_{n \to \infty} \int_E f_n(x) dx.$$

1.7 Exercises

1.1. Show that $\int_0^1 x^{m-1}(1-x)^{n-1}dx = \dfrac{\Gamma(m)\Gamma(n)}{\Gamma(m+n)}$ for $m > 0, n > 0$.

1.2. Show that (i) $E_2(-x^2) = \cos x$ (ii) $E_2(x^2) = \cosh x$ (iii) $x E_{2,2}(x^2) = \sinh x$.

1.3. Show that $\frac{d}{dx}E_{\alpha,\beta}(x) = \frac{d}{dx}E_{\alpha,\alpha+\beta}(x) + E_{\alpha,\alpha+\beta}(x)$.

1.4. Calculate $E_\alpha(At^\alpha)$ when (i) $A = \begin{bmatrix} 1 & -1 \\ 1 & 1 \end{bmatrix}$, (ii) $A = \begin{bmatrix} 0 & -1 \\ 1 & 0 \end{bmatrix}$.

1.5. Find the Laplace transform of (i) $\cos at$, (ii) e^{at}, (iii) $\dfrac{\sin at \sinh at}{2a^2}$.

1.6. Find the inverse Laplace transform of (i) $\dfrac{1}{(s-2)^2}$, (ii) $\dfrac{1}{s^n}$, (iii) $\dfrac{s}{s^2+a^2}$.

1.7. Prove the Banach fixed point theorem.

1.8. Define $T : C[0, 1] \to C[0, 1]$ by $Tx(t) = 1 + \int_0^1 x(s)ds$. Is T a contraction?

1.9. Show that the operator $T : C[0, 1] \to C[0, 1]$ defined by $Tx(t) = \int_0^t (t - s)x(s)ds$ is a contraction.

1.10. Show by an example that Brouwer's theorem fails in infinite-dimensional spaces.

1.11. Give an example of bounded operator.

1.12. Give an example of completely continuous operator.

References

1. Miller, K., Ross, B.: An Introduction to the Fractional Calculus and Fractional Differential Equations. John Wiley and Sons Inc, New York (1993)
2. Capelas de Oliveira, E.: Solved Exercises in Fractional Calculus. Springer Nature, Switzerland (2019)
3. Popolizio, M.: On the matrix Mittlag-Leffler function. Theoretical properties and numerical computation. Mathematics 7(12), 1140 (2019)
4. Smart, D.R.: Fixed Point Theorems. Cambridge University Press, Cambridge (1980)

Chapter 2
Fractional Calculus

Abstract After introducing the concept of fractional integral we define the Riemann–Liouville fractional integrals, Riemann–Liouville fractional derivatives, and Caputo fractional derivatives. Some elementary properties are proved and few examples are added. In the end a set of exercises is included.

Keywords Fractional calculus · Riemann–Liouville fractional integral · Riemann–Liouville fractional derivative · Caputo fractional derivative · Properties and examples

2.1 Preliminaries

Let $f(t)$ be a function defined for $t > 0$ and define the definite integral from 0 to t as

$$If(t) = \int_0^t f(s)\mathrm{d}s.$$

Repeating this process

$$I^2 f(t) = \int_0^t If(s)\mathrm{d}s = \int_0^t \left(\int_0^s f(\tau)\mathrm{d}\tau \right) \mathrm{d}s,$$

and this can be extended arbitrarily. The Cauchy formula for repeated integration, namely,

$$I^n f(t) = \frac{1}{(n-1)!} \int_0^t (t-s)^{n-1} f(s)\mathrm{d}s$$

leads in a straightforward way to a generalization for real n.

Using the gamma function to remove the discrete nature of the factorial function gives us a natural candidate for fractional applications of the integral operator

© The Author(s), under exclusive license to Springer Nature Singapore Pte Ltd. 2023
K. Balachandran, *An Introduction to Fractional Differential Equations*, Industrial and Applied Mathematics, https://doi.org/10.1007/978-981-99-6080-4_2

$$I^\alpha f(t) = \frac{1}{\Gamma(\alpha)} \int_0^t (t-s)^{\alpha-1} f(s) ds,$$

where $\alpha > 0$. This is a well-defined operator. It is straightforward to show that the operator I satisfies

$$I^\alpha I^\beta f(t) = I^\beta I^\alpha f(t) = I^{\alpha+\beta} f(t) = \frac{1}{\Gamma(\alpha+\beta)} \int_0^t (t-s)^{\alpha+\beta-1} f(s) ds.$$

In 1832, J. Liouville gave two definitions of fractional derivatives of a fairly restrictive class of functions. The first definition restricts functions to functions that can be expressed as a trigonometric series and the second definition applies to the functions $x^{-\alpha}$ with $\alpha > 0$. The definition of the fractional integral begins with the consideration of the n-fold integral

$$I_a^n f(x) = \int_a^x dx_1 \int_a^{x_1} dx_2 \int_a^{x_2} dx_3 \cdots \int_a^{x_{n-1}} f(t) dt. \tag{2.1}$$

The function f in (2.1) is assumed to be continuous on the interval $[a, b]$ where $b > a$. We assert that (2.1) may be reduced to a single integral of the form

$$I_a^n f(x) = \int_a^x \frac{(x-t)^{n-1}}{(n-1)!} f(t) dt = \frac{1}{\Gamma(n)} \int_a^x (x-t)^{n-1} f(t) dt. \tag{2.2}$$

Clearly the right-hand side of expression (2.2) is meaningful for any number n whose real part is greater than zero, that is,

$$I_a^\alpha f(x) = \frac{1}{\Gamma(\alpha)} \int_a^x (x-t)^{\alpha-1} f(t) dt, \tag{2.3}$$

where $\mathrm{Re}\, \alpha > 0$, which is the fractional integral of f of order α.

Alternatively consider the n-th-order differential equation

$$\frac{d^n y}{dx^n} = f(x)$$

with initial conditions

$$y(a) = y'(a) = y''(a) = \cdots = y^{n-1}(a) = 0.$$

Then the solution of the equation is

$$y(x) = \int_a^x \frac{(x-t)^{n-1}}{(n-1)!} f(t) dt.$$

Since $f(x)$ is the n-th derivative of $y(x)$, we may interpret $y(x)$ as the n-th integral of $f(x)$. Thus

$$I_a^n f(x) = \int_a^x \frac{(x-t)^{n-1}}{(n-1)!} f(t) dt.$$

If we replace integer n with any real number α and change the factorial to a gamma function, then we have (2.3).

Another way of defining the fractional integral is as follows.

We can recognize the Laplace domain equivalent for the n-fold integral of the function $f(t)$. Consider an anti-derivative or primitive of the function $f(t)$, $\mathcal{D}^{-1} f(t)$, then

$$\mathcal{D}^{-1} f(t) = \int_0^t f(x) dx. \tag{2.4}$$

Now let us perform the repeated applications of the operator. For example,

$$\mathcal{D}^{-2} f(t) = \int_0^t \int_0^x f(y) dy dx. \tag{2.5}$$

Equation (2.5) can be considered as a double integral and taking into account the xy-plane over which it is integrated (see Fig. 2.1), we can reverse the sequence of integrations by doing the proper changes in their limits. So we obtain

$$\mathcal{D}^{-2} f(t) = \int_0^t \int_y^t f(y) dx dy. \tag{2.6}$$

As $f(y)$ is a constant with respect to x, we find that the inner integral is simply $(t-y) f(y)$ and we have

$$\mathcal{D}^{-2} f(t) = \int_0^t (t-y) f(y) dy. \tag{2.7}$$

Similarly we can obtain

$$\mathcal{D}^{-3} f(t) = \frac{1}{2} \int_0^t (t-y)^2 f(y) dy, \tag{2.8}$$

and so on giving the formula

$$\mathcal{D}^{-n} f(t) = \underbrace{\int \cdots \int_0^t}_{n} f(y) \underbrace{dy \cdots dy}_{n} = \int_0^t \frac{f(y)(t-y)^{n-1}}{(n-1)!} dy. \tag{2.9}$$

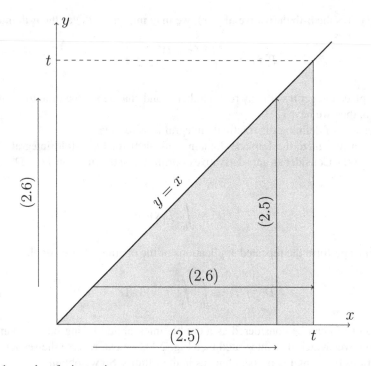

Fig. 2.1 xy-plane for integration

Since $\mathcal{D}^{-n} = I^n$, Eq. (2.9) can be written as

$$I^n f(t) = \frac{1}{(n-1)!} \int_0^t (t-s)^{n-1} f(s)\mathrm{d}s.$$

This is true for any $n > 0$ and so we replace n by α for fractional nature and factorial by gamma and have (2.3) for $a = 0$.

2.2 Riemann–Liouville Fractional Integrals and Derivatives

Let $J = [a, b]$ $(-\infty < a < b < \infty)$ be a finite interval on the real line $\mathbb{R}$. The Riemann–Liouville fractional integrals $I_{a+}^\alpha f(x)$ and $I_{b-}^\alpha f(x)$ of order $\alpha \in \mathbb{C}$ $(\mathbb{R}(\alpha) > 0)$ are defined by

$$I_{a+}^\alpha f(x) := \frac{1}{\Gamma(\alpha)} \int_a^x (x-t)^{\alpha-1} f(t)\mathrm{d}t, \ x > a, \ \mathbb{R}(\alpha) > 0 \qquad (2.10)$$

and

$$I_{b-}^\alpha f(x) := \frac{1}{\Gamma(\alpha)} \int_x^b (t-x)^{\alpha-1} f(t) dt, \ x < b, \ \mathbb{R}(\alpha) > 0. \tag{2.11}$$

These integrals are called the left-sided and the right-sided fractional integrals. The Riemann–Liouville fractional derivatives $D_{a+}^\alpha f(x)$ and $D_{b-}^\alpha f(x)$ of order $\alpha \in \mathbb{C}$ ($\mathbb{R}(\alpha) \geq 0$) are defined by [1]

$$D_{a+}^\alpha f(x) = D^n I_{a+}^{n-\alpha} f(x), \tag{2.12}$$

$$= \frac{1}{\Gamma(n-\alpha)} \frac{d^n}{dx^n} \int_a^x \frac{f(t)}{(x-t)^{\alpha-n+1}} dt, \ n = [\mathbb{R}(\alpha)] + 1, x > a,$$

and

$$D_{b-}^\alpha f(x) = (-\frac{d^n}{dx^n}) I_{0+}^{n-\alpha} f(x), \tag{2.13}$$

$$= \frac{1}{\Gamma(n-\alpha)} (-\frac{d^n}{dx^n}) \int_x^b \frac{f(t)}{(t-x)^{\alpha-n+1}} dt, \ n = [\mathbb{R}(\alpha)] + 1, x < b,$$

where $[\mathbb{R}(\alpha)]$ means the integral part of $\mathbb{R}(\alpha)$ and $D = \frac{d}{dx}$. When $\alpha = n \in N_0 = \{0, 1, 2, \cdots\}$,

$$D_{a+}^n f(x) = D^m (I_{a+}^{m-n}) f(x), \ m = [\mathbb{R}(n)] + 1.$$

Suppose $n = 0$, then

$$D_{a+}^0 f(x) = D^m \left(I_{a+}^m f(x) \right), \ m = [\mathbb{R}(0)] + 1,$$

$$= \frac{d}{dx} (I_{a+}^1 f(x)),$$

$$= \frac{d}{dx} \int_a^x f(t) dt.$$

By using Leibnitz formula, we get

$$D_{a+}^0 f(x) = f(x).$$

Similarly when $n = 1, n = 2, n = 3, \cdots$, we get

$$D_{a+}^1 f(x) = f^1(x),$$
$$D_{a+}^2 f(x) = f^2(x),$$
$$D_{a+}^3 f(x) = f^3(x),$$

$$\vdots$$

$$D_{a+}^n f(x) = f^n(x).$$

If $0 < \mathbb{R}(\alpha) < 1$, then (2.12) and (2.13) become

$$D_{a+}^{\alpha} f(x) = \frac{1}{\Gamma(1-\alpha)} \frac{d}{dx} \int_a^x \frac{f(t)}{(x-t)^{\alpha-[\mathbb{R}(\alpha)]}} dt,$$

for $0 < \mathbb{R}(\alpha) < 1, x > a$ and

$$D_{b-}^{\alpha} f(x) = \frac{-1}{\Gamma(1-\alpha)} \frac{d}{dx} \int_x^b \frac{f(t)}{(t-x)^{\alpha-[\mathbb{R}(\alpha)]}} dt,$$

for $0 < \mathbb{R}(\alpha) < 1, x < b$, respectively. When $\alpha \in \mathbb{R}^+$, (2.12) and (2.13) become

$$D_{a+}^{\alpha} f(x) = \frac{1}{\Gamma(n-\alpha)} \frac{d^n}{dx^n} \int_a^x \frac{f(t)}{(x-t)^{\alpha-n+1}} dt,$$

for $n = [\alpha] + 1, x > a$, and

$$D_{b-}^{\alpha} f(x) = \frac{1}{\Gamma(n-\alpha)} (-\frac{d^n}{dx^n}) \int_x^b \frac{f(t)}{(t-x)^{\alpha-n+1}} dt,$$

for $n = [\alpha] + 1, x < b$, where $0 < \mathbb{R}(\alpha) < 1$ implies $\alpha \in \mathbb{R}^+$

$$D_{a+}^{\alpha} f(x) = \frac{1}{\Gamma(1-\alpha)} \frac{d}{dx} \int_a^x \frac{f(t)}{(x-t)^{\alpha}} dt, \quad 0 < \alpha < 1, x > a,$$

and

$$D_{b-}^{\alpha} f(x) = \frac{-1}{\Gamma(n-\alpha)} \frac{d}{dx} \int_x^b \frac{f(t)}{(t-x)^{\alpha}} dt, \quad 0 < \alpha < 1, t < b.$$

If $\mathbb{R}(\alpha) = 0$ but $(\alpha \neq 0)$, (2.12) and (2.13) become the fractional derivatives of a purely imaginary order.

$$D_{a+}^{i\theta} f(x) = \frac{1}{\Gamma(1-i\theta)} \frac{d}{dx} \int_a^x \frac{f(t)}{(x-t)^{i\theta}} dt, \quad \theta \in \mathbb{R}/\{0\}, x > a,$$

and

$$D_{b-}^{i\theta} f(x) = \frac{1}{\Gamma(1-i\theta)} \frac{d}{dx} \int_x^b \frac{f(t)}{(t-x)^{i\theta}} dt, \quad \theta \in \mathbb{R}/\{0\}, t < b.$$

Theorem 2.2.1 *If* $\mathbb{R}(\alpha) \geq 0$ *and* $\beta \in \mathbb{C}, (\mathbb{R}(\beta) > 0)$, *then*

$$I_{a+}^{\alpha} (x-a)^{\beta-1} = \frac{\Gamma(\beta)}{\Gamma(\beta+\alpha)} (x-a)^{\beta+\alpha-1}, \mathbb{R}(\alpha) > 0;$$

$$D_{a+}^{\alpha} (x-a)^{\beta-1} = \frac{\Gamma(\beta)}{\Gamma(\beta-\alpha)} (x-a)^{\beta-\alpha-1}, \mathbb{R}(\alpha) \geq 0.$$

Proof The Riemann–Liouville integral (2.10) can be written in the form

$$I_{a+}^{\alpha} f(x) = \frac{1}{\Gamma(\alpha)} \int_a^x \frac{f(t)}{(x-t)^{1-\alpha}} dt, \, x > a, \mathbb{R}(\alpha) > 0.$$

Taking $f(t) = (t-a)^{\beta-1}$,

$$I_{a+}^{\alpha}(x-a)^{\beta-1} = \frac{1}{\Gamma(\alpha)} \int_a^x \frac{(t-a)^{\beta-1}}{(x-t)^{1-\alpha}} dt,$$
$$= \frac{1}{\Gamma(\alpha)} \int_a^x (t-a)^{\beta-1}(x-t)^{\alpha-1} dt.$$

Substituting the transformation

$$y = \frac{t-a}{x-a}, dy = \frac{dt}{x-a} \implies dt = (x-a)dy,$$

we get

$$I_{a+}^{\alpha}(x-a)^{\beta-1} = \frac{1}{\Gamma(\alpha)} \int_a^x (t-a)^{\beta-1}(x-t)^{\alpha-1} dt,$$
$$= \frac{(x-a)^{\beta+\alpha-1}}{\Gamma(\alpha)} \int_0^1 y^{\beta-1}(1-y)^{\alpha-1} dy,$$
$$= \frac{\Gamma(\beta)}{\Gamma(\alpha+\beta)}(x-a)^{\beta+\alpha-1}.$$

Thus we have

$$I_{a+}^{\alpha}(x-a)^{\beta-1} = \frac{\Gamma(\beta)}{\Gamma(\alpha+\beta)}(x-a)^{\beta+\alpha-1}, \, \mathbb{R}(\alpha) > 0.$$

Now the Riemann–Liouville derivative (2.12) can be written as

$$D_{a+}^{\alpha}(f(x)) = D^n(I_{a+}^{n-\alpha} f(x)).$$

Taking $f(t) = (t-a)^{\beta-1}$, we have

$$D_{a+}^{\alpha}(x-a)^{\beta-1} = D^n(I_{a+}^{n-\alpha}(x-a)^{\beta-1})$$
$$= \frac{1}{\Gamma(n-\alpha)} \frac{d^n}{dx^n} \int_a^x (t-a)^{\beta-1}(x-t)^{n-\alpha-1} dt$$
$$= \frac{1}{(-(\alpha-1))!} \int_a^x (t-a)^{\beta-1}(x-t)^{-\alpha-1} dt.$$

Using the bilinear transformation

$$y = \frac{t-a}{x-a}, dy = \frac{dt}{x-a} \implies dt = (x-a)dy,$$

we get

$$D_{a+}^{\alpha}(x-a)^{\beta-1} = \frac{(x-a)^{\beta-\alpha-1}}{(-(\alpha-1))!} \int_0^1 y^{\beta-1}(1-y)^{-\alpha-1}dy,$$

$$= \frac{\Gamma(\beta)}{\Gamma(\beta-\alpha)}(x-a)^{\beta-\alpha-1}, \mathbb{R}(\alpha) \geq 0.$$

□

Similar to the above theorem, we can also prove

$$I_{b-}^{\alpha}(b-x)^{\beta-1} = \frac{\Gamma(\beta)}{\Gamma(\beta+\alpha)}(b-x)^{\beta+\alpha-1}, \mathbb{R}(\alpha) > 0;$$

$$D_{b-}^{\alpha}(b-x)^{\beta-1} = \frac{\Gamma(\beta)}{\Gamma(\beta-\alpha)}(b-x)^{\beta-\alpha-1}, \mathbb{R}(\alpha) \geq 0.$$

In particular, if $\beta = 1$ and $\mathbb{R}(\alpha) \geq 0$, then the Riemann–Liouville fractional derivatives of a constant are, in general, not equal to zero. We know that

$$D_{a+}^{\alpha}(x-a)^{\beta-1} = \frac{\Gamma(\beta)}{\Gamma(\beta-\alpha)}(x-a)^{\beta-\alpha-1},$$

$$D_{a+}^{\alpha}1 = \frac{\Gamma(1)}{\Gamma(1-\alpha)}(x-a)^{-\alpha} = \frac{(x-a)^{-\alpha}}{\Gamma(1-\alpha)}$$

$$D_{b-}^{\alpha}1 = \frac{(b-x)^{-\alpha}}{\Gamma(1-\alpha)}.$$

Lemma 2.2.2 *If* $\mathbb{R}(\alpha) > 0$ *and* $\mathbb{R}(\beta) > 0$, *then the equations*

$$I_{a+}^{\alpha}I_{a+}^{\beta}f(x) = I_{a+}^{\alpha+\beta}f(x)$$

and

$$I_{b-}^{\alpha}I_{b-}^{\beta}f(x) = I_{b-}^{\alpha+\beta}f(x)$$

are satisfied at almost every point $x \in [a, b]$ *for* $f(x) \in L_p(a, b)$, $1 \leq p \leq \infty$. *If* $\alpha + \beta > 1$, *then the relations hold at any point of* $[a, b]$.

Proof By definition, we have

$$I^\alpha_{a+}(I^\beta_{a+}f(x)) = \frac{1}{\Gamma(\alpha)} \int_a^x (x-s)^{\alpha-1}(I^\beta_{a+}f(s))ds$$

$$= \frac{1}{\Gamma(\alpha)\Gamma(\beta)} \int_a^x \int_a^s (x-s)^{\alpha-1}(s-\tau)^{\beta-1}f(\tau)d\tau ds$$

$$= \frac{1}{\Gamma(\alpha)\Gamma(\beta)} \int_a^x f(\tau)\left(\int_\tau^x (x-s)^{\alpha-1}(s-\tau)^{\beta-1}ds \right)d\tau$$

where in the last step we exchanged the order of integration and pulled out the $f(\tau)$ factor from the s integration. Changing variables to r defined by $s = \tau + (x - \tau)r$,

$$I^\alpha_{a+}(I^\beta_{a+}f(x)) = \frac{1}{\Gamma(\alpha)\Gamma(\beta)} \int_a^x (x-\tau)^{\alpha+\beta-1}f(\tau)\left(\int_0^1 (1-r)^{\alpha-1}r^{\beta-1}dr \right)d\tau.$$

The inner integral is the beta function which satisfies the following property:

$$\int_0^1 (1-r)^{\alpha-1}r^{\beta-1}dr = B(\alpha, \beta) = \frac{\Gamma(\alpha)\Gamma(\beta)}{\Gamma(\alpha+\beta)}.$$

Substituting back into the equation

$$I^\alpha_{a+}(I^\beta_{a+}f(x)) = \frac{1}{\Gamma(\alpha+\beta)} \int_a^x (x-\tau)^{\alpha+\beta-1}f(\tau)d\tau = I^{\alpha+\beta}_{a+}f(x),$$

and interchanging α and β show that the order in which the operator I^α_{a+} is applied is immaterial. By similar way the other identity can be proved. □

This relationship is called the semigroup property of fractional integral operators. Unfortunately the comparable process for the fractional derivative operator D^α_a is significantly more complex, but it can be shown that D^α_a is neither commutative nor additive in general.

Theorem 2.2.3 *If* $\mathbb{R}(\alpha) > \mathbb{R}(\beta) > 0$, *then for* $f(x) \in L_p(a, b)$, $1 \leq p \leq \infty$, *the relations*

$$D^\beta_{a+}I^\alpha_{a+}f(x) = I^{\alpha-\beta}_{a+}f(x)$$

and

$$D^\beta_{b-}I^\alpha_{b-}f(x) = I^{\alpha-\beta}_{b-}f(x)$$

hold almost everywhere on $[a, b]$.

Proof From the Riemann–Liouville integral (2.10) and derivative (2.12), we have

$$D_{a+}^\beta I_{a+}^\alpha f(x) = D^n I_{a+}^{n+\alpha-\beta} f(x).$$

Using the above lemma, we have

$$D_{a+}^\beta I_{a+}^\alpha f(x) = \frac{1}{\Gamma(n+\alpha-\beta)} \left(\frac{d^n}{dx^n}\right) \int_a^x \frac{f(t)}{(x-t)^{\beta-\alpha-n+1}} dt.$$

By Leibniz formula, we have

$$
\begin{aligned}
D_{a+}^\beta I_{a+}^\alpha f(x) &= \frac{1}{\Gamma(n+\alpha-\beta)} \int_a^x \frac{\Gamma(n+\alpha-\beta)}{\Gamma(\alpha-\beta)}(x-t)^{\alpha-\beta-1} f(t)dt, \\
&= \frac{1}{\Gamma(\alpha-\beta)} \int_a^x \frac{f(t)}{(x-t)^{\beta-\alpha+1}} dt, \\
&= I_{a+}^{\alpha-\beta} f(x);
\end{aligned}
$$

hence

$$D_{a+}^\beta I_{a+}^\alpha f(x) = I_{a+}^{\alpha-\beta} f(x).$$

Similarly

$$D_{b-}^\beta I_{b-}^\alpha f(x) = I_{b-}^{\alpha-\beta} f(x).$$

□

Observe that when $\beta = 1$ we have

$$DI_{a+}^{\alpha+1} f(x) = I_{a+}^\alpha f(x).$$

Theorem 2.2.4 *Let* $\mathbb{R}(\alpha) \geq 0$, $m \in \mathbb{N}$.

(a) *If the fractional derivatives* $D_{a+}^\alpha f(x)$ *and* $D_{a+}^{\alpha+m} f(x)$ *exist, then*

$$D^m D_{a+}^\alpha f(x) = D_{a+}^{\alpha+m} f(x).$$

(b) *If the fractional derivatives* $D_{b-}^\alpha f(x)$ *and* $(D_{b-}^{\alpha+m} f)(x)$ *exist, then*

$$D^m D_{b-}^\alpha f(x) = D_{a+}^{\alpha+m} f(x).$$

Proof By using the previous theorem for $\beta = k \in \mathbb{N}$ and $\mathbb{R}(\alpha) > k$, we have

$$D^k I_{a+}^\alpha f(x) = I_{a+}^{\alpha-k} f(x)$$

and

$$D^k I_{b-}^{\alpha} f(x) = I_{b-}^{\alpha-k} f(x).$$

If the Riemann–Liouville derivative can be written in the form

$$D_{a+}^{\alpha} f(x) = D^n I_{a+}^{n-\alpha} f(x),$$

then

$$
\begin{aligned}
D^m D_{a+}^{\alpha} f(x) &= D^m D^n I_{a+}^{n-\alpha} f(x) \\
&= D^n D^m I_{a+}^{n-\alpha} f(x) \\
&= D^n I_{a+}^{n-\alpha-m} f(x) \\
&= D_{a+}^{\alpha+m} f(x).
\end{aligned}
$$

Therefore

$$D^m D_{a+}^{\alpha} f(x) = D_{a+}^{\alpha+m} f(x).$$

Similarly

$$D^m D_{b-}^{\alpha} f(x) = D_{b-}^{\alpha+m} f(x).$$

$\square$

Lemma 2.2.5 *If* $\mathbb{R}(\alpha) > 0$ *and* $f(x) \in L_p(a, b)$, $1 \le p \le \infty$, *then the following equalities hold:*

$$D_{a+}^{\alpha} I_{a+}^{\alpha} f(x) = f(x)$$

and

$$D_{b-}^{\alpha} I_{b-}^{\alpha} f(x) = f(x).$$

Proof The Riemann–Liouville integral (2.10) and derivative (2.12) can be written in the form

$$I_{a+}^{\alpha} f(x) = \frac{1}{\Gamma(\alpha)} \int_a^x \frac{f(t)}{(x-t)^{1-\alpha}} dt$$

and

$$D_{a+}^{\alpha} f(x) = D^n I_{a+}^{n-\alpha} f(x).$$

Then

$$D_{a+}^{\alpha} I_{a+}^{\alpha} f(x) = D^n I_{a+}^n f(x).$$

Take $D^n(I_{a+}^n f)(x)$; when $n = 1$

$$\frac{d}{dx} I_{a+}^1 f(x) = \frac{d}{dx} \frac{1}{\Gamma(1)} \int_a^x f(t) dt.$$

By using Leibniz rule, we have

$$\frac{d}{dx} I_{a+}^1 f(x) = f(x).$$

When $n = 2$

$$\frac{d^2}{dx^2} I_{a+}^2 f(x) = f(x).$$

By proceeding in a similar manner, we get

$$D^n I_{a+}^n f(x) = f(x).$$

Thus

$$D_{a+}^{\alpha} I_{a+}^{\alpha} f(x) = f(x).$$

Similarly

$$D_{b-}^{\alpha} I_{b-}^{\alpha} f(x) = f(x).$$

$\square$

Note that $I_{a+}^{\alpha} D_{a+}^{\alpha} f(x)$ is not necessarily equal to $f(x)$ but the equality holds when $f(x) \in I_{a+}^{\alpha}(L_1)$, that is, $I_{a+}^{\alpha} D_{a+}^{\alpha} f(x) = f(x)$. Instead, if $f(x) \in L_1(a, b)$ has an integrable fractional derivative $D_{a+}^{\alpha} f(x)$ then the equality is not true in general [2].

Laplace Transform of Riemann–Liouville Fractional Derivative

The Laplace transform of the Riemann–Liouville integral of order $\alpha > 0$ of a function $f(x)$ can be written as a convolution of the functions $\frac{1}{\Gamma(\alpha)} x^{\alpha-1}$ and $f(x)$:

$$I_{0+}^{\alpha} f(x) = \frac{1}{\Gamma(\alpha)} \int_0^x (x - t)^{\alpha-1} f(t) dt = \frac{1}{\Gamma(\alpha)} x^{\alpha-1} * f(x).$$

The Laplace transform of the function $x^{\alpha-1}$ is

$$\mathcal{L}[x^{\alpha-1}](s) = \Gamma(\alpha)s^{-\alpha}.$$

Here we use the formula

$$\mathcal{L}[I_{0+}^{\alpha} f(x)](s) = \mathcal{L}(\frac{1}{\Gamma(\alpha)}x^{\alpha-1})\mathcal{L}(f(x)) = s^{-\alpha}F(s). \tag{2.14}$$

To obtain the Laplace transform of Riemann–Liouville derivative, we represent $D_{0+}^{\alpha} f(x)$ as the n-th derivative of some function, say $g(x)$. That is,

$$D_{0+}^{\alpha} f(x) = g^{(n)}(x),$$

and so $g(x)$ can be written as

$$g(x) = D_{0+}^{-(n-\alpha)} f(x)$$
$$= \frac{1}{\Gamma(n-\alpha)} \int_0^x (x-t)^{n-\alpha-1} f(t)dt, \ n-1 \le \alpha < n.$$

The use of the formula for the Laplace transform of an integer order derivative leads to

$$\mathcal{L}[D_{0+}^{\alpha} f(x)](s) = \mathcal{L}[g^{(n)}(x)](s) = s^n G(s) - \sum_{k=0}^{n-1} s^k g^{(n-k-1)}(0). \tag{2.15}$$

Using (2.14), the Laplace transform of $g(x)$ is evaluated as

$$G(s) = s^{-(n-\alpha)} F(s). \tag{2.16}$$

Additionally, from the definition of the Riemann–Liouville fractional derivative,

$$g^{(n-k-1)}(x) = \frac{d^{n-k-1}}{dx^{n-k-1}} D_{0+}^{-(n-\alpha)} f(x),$$
$$= D_{0+}^{\alpha-k-1} f(x). \tag{2.17}$$

Substituting (2.16) and (2.17) in (2.15) we obtain the Laplace transform of the Riemann–Liouville derivative of order $\alpha > 0$ as

$$\mathcal{L}[D_{0+}^{\alpha} f(x)](s) = s^{\alpha} F(s) - \sum_{k=0}^{n-1} s^k D_{0+}^{\alpha-k-1} f(x)|_{x=0}, \ n-1 \le \alpha < n.$$

2.3 Caputo Fractional Derivatives

Let $D_{a+}^{\alpha} f(x)$, $D_{b-}^{\alpha} f(x)$ be the Riemann–Liouville fractional derivatives of order $\alpha \in \mathbb{C}$ ($\mathbb{R}(\alpha) \geq 0$) defined by (2.12) and (2.13). The fractional derivatives $^C D_{a+}^{\alpha} f(x)$ and $^C D_{b-}^{\alpha} f(x)$ of order $\alpha \in \mathbb{C}$ ($\mathbb{R}(\alpha) \geq 0$) on $[a, b]$ are defined via the above Riemann–Liouville fractional derivatives by

$$^C D_{a+}^{\alpha} f(x) = D_{a+}^{\alpha} [f(x) - \sum_{k=0}^{n-1} \frac{f^{(k)}(a)}{k!} (x - a)^k],$$

$$^C D_{b-}^{\alpha} f(x) = D_{b-}^{\alpha} [f(x) - \sum_{k=0}^{n-1} \frac{f^{(k)}(b)}{k!} (b - x)^k],$$

respectively, where $n = [\mathbb{R}(\alpha)] + 1$ for $\alpha \notin \mathbb{N}_0$, $n = \alpha \in \mathbb{N}_0$. These derivatives are called left-sided and right-sided Caputo fractional derivatives of order α. When $0 < \mathbb{R}(\alpha) < 1$, the relations take the following forms:

$$^C D_{a+}^{\alpha} f(x) = D_{a+}^{\alpha} [f(x) - f(a)],$$
$$^C D_{b-}^{\alpha} f(x) = D_{b-}^{\alpha} [f(x) - f(b)].$$

If $\alpha \notin \mathbb{N}_0$ and $f(x)$ is a function for which the Caputo fractional derivatives $^C D_{a+}^{\alpha} f(x)$ and $^C D_{b-}^{\alpha} f(x)$ of order $\alpha \in \mathbb{C}$, ($\mathbb{R}(\alpha) \geq 0$) exist together with the Riemann–Liouville fractional derivatives $(D_{a+}^{\alpha} f)(x)$ and $(D_{b-}^{\alpha} f)(x)$, then

$$D_{a+}^{0} f(x) = D_{b-}^{0} f(x) = f(x),$$
$$D_{a+}^{n} f(x) = f^{(n)}(x),$$
$$D_{b-}^{n} f(x) = (-1)^n f^{(n)}(x), \quad (n \in \mathbb{N})$$

and

$$D_{a+}^{\alpha} (x - a)^{\beta-1} = \frac{\Gamma(\beta)}{\Gamma(\beta - \alpha)} (x - a)^{\beta-\alpha-1}, \quad (\mathbb{R}(\alpha) \geq 0).$$

They are connected with each other by the following relations:

$$^C D_{a+}^{\alpha} f(x) = D_{a+}^{\alpha} f(x) - \sum_{k=0}^{n-1} \frac{f^k(a)}{k!} D_{a+}^{\alpha} (x - a)^k$$

$$= D_{a+}^{\alpha} f(x) - \sum_{k=0}^{n-1} \frac{f^k(a)}{k!} \frac{\Gamma(k+1)}{\Gamma(k - \alpha + 1)} (x - a)^{(k-\alpha)}$$

$$= D_{a+}^{\alpha} f(x) - \sum_{k=0}^{n-1} \frac{f^k(a)}{\Gamma(k - \alpha + 1)} (x - a)^{k-\alpha}, \quad n = [\mathbb{R}(\alpha)] + 1,$$

and similarly

$$^C D_{b-}^\alpha f(x) = D_{b-}^\alpha f(x) - \sum_{k=0}^{n-1} \frac{f^{(k)}(b)}{\Gamma(k-\alpha+1)} (b-x)^k, \; n = [\mathbb{R}(\alpha)] + 1.$$

When $0 < \mathbb{R}(\alpha) < 1$, we have

$$^C D_{a+}^\alpha f(x) = D_{a+}^\alpha f(x) - \frac{f(a)}{\Gamma(1-\alpha)} (x-a)^{-\alpha}$$

$$^C D_{b-}^\alpha f(x) = D_{b-}^\alpha f(x) - \frac{f(b)}{\Gamma(1-\alpha)} (b-x)^{-\alpha}.$$

If $\alpha \notin \mathbb{N}_0$, then the Caputo fractional derivatives coincide with the Riemann–Liouville fractional derivatives in the following case:

$$^C D_{a+}^\alpha f(x) = D_{a+}^\alpha f(x),$$

if $f(a) = f^1(a) = \cdots = f^{n-1}(a) = 0, n = [\mathbb{R}(\alpha)] + 1$ and

$$^C D_{b-}^\alpha f(x) = D_{b-}^\alpha f(x),$$

if $f(b) = f^1(b) = \cdots = f^{n-1}(b) = 0, \; n = [\mathbb{R}(\alpha)] + 1$. In particular, when $0 < \mathbb{R}(\alpha) < 1$,

$$^C D_{a+}^\alpha f(x) = D_{a+}^\alpha f(x), \text{ when } f(a) = 0,$$
$$^C D_{b-}^\alpha f(x) = D_{b-}^\alpha f(x), \text{ when } f(b) = 0.$$

If $\alpha = n \in \mathbb{N}_0$, $f^n(x)$ is the usual derivative of $f(x)$ of order $n, n \in \mathbb{N}$, and

$$^C D_{a+}^\alpha f(x) = f^n(x),$$
$$^C D_{b-}^\alpha f(x) = (-1)^n f^n(x).$$

The left-sided and right-sided Caputo fractional derivatives of order $\alpha > 0$, $n - 1 < \alpha \le n$, are defined by [3]

$$^C D_{a+}^\alpha f(x) := I_{a+}^{n-\alpha} D^n f(x)$$

$$= \frac{1}{\Gamma(n-\alpha)} \int_a^x (x-t)^{n-\alpha-1} f^{(n)}(t) dt, \qquad (2.18)$$

and

$$^C D_{b-}^\alpha f(x) := (-1)^n I_{0+}^{n-\alpha} D^n f(x)$$

$$= \frac{(-1)^n}{\Gamma(n-\alpha)} \int_x^b (t-x)^{n-\alpha-1} f^{(n)}(t) dt, \qquad (2.19)$$

respectively, where the function $f(t)$ has absolutely continuous derivatives up to order $(n-1)$ on $[a, b]$. In particular, when $0 < \alpha < 1$,

$$^CD_{a+}^\alpha f(x) = \frac{1}{\Gamma(1-\alpha)} \int_a^x (x-t)^{-\alpha} f'(t)dt, \tag{2.20}$$

and

$$^CD_{b-}^\alpha f(t) = \frac{-1}{\Gamma(1-\alpha)} \int_x^b (t-x)^{-\alpha} f'(t)dt. \tag{2.21}$$

Laplace Transform of Caputo Derivative

The Laplace transform of Caputo derivative is obtained by writing the Caputo derivative in the form

$$^CD_{0+}^\alpha f(x) = D_{0+}^{-(n-\alpha)} g(x), \ g(x) = f^{(n)}(x).$$

By using (2.14) we have

$$\mathcal{L}[D_{0+}^{-(n-\alpha)} g(x)](s) = s^{-(n-\alpha)} G(s). \tag{2.22}$$

The Laplace transform of an integer order derivative is

$$G(s) = s^n F(s) - \sum_{k=0}^{n-1} s^{n-k-1} f^{(k)}(0). \tag{2.23}$$

Substituting (2.23) in (2.22), we have

$$\mathcal{L}[^CD_{0+}^\alpha f(x)](s) = s^\alpha F(s) - \sum_{k=0}^{n-1} s^{\alpha-k-1} f^{(k)}(0), \ n-1 < \alpha \le n.$$

Notations: For brevity and notational convenience, here afterward, we take the Riemann–Liouville fractional integral as I^α for I_0^α, Riemann–Liouville fractional derivative D^α for D_0^α and the Caputo derivative $^CD_0^\alpha$ as $^CD^\alpha$. Even though many definitions of fractional integrals and derivatives are introduced by several authors (a partial list is given in Chap. 6) we are considering in this book only Caputo derivative and R-L integral and derivative.

The relation between the Caputo and Riemann–Liouville fractional derivatives is proved in the following theorem.

Theorem 2.3.1 *Suppose $x > 0$, $\alpha \in \mathbb{R}$ and $n-1 < \alpha < n$, $n \in N$. Then the following relation between the Riemann–Liouville and the Caputo derivatives holds:*

$$^{C}D^{\alpha}f(x) = D^{\alpha}f(x) - \sum_{k=0}^{n-1} \frac{x^{k-\alpha}}{\Gamma(k+1-\alpha)} f^{(k)}(0).$$

Proof The well-known Taylor series expansion of $f(x)$ about the point 0 is

$$f(x) = f(0) + xf'(0) + \frac{x^2}{2!}f''(0) + \frac{x^3}{3!}f'''(0) + \cdots + \frac{x^{n-1}}{(n-1)!}f^{n-1}(0) + R_{n-1}$$

$$= \sum_{k=0}^{n-1} \frac{x^k}{\Gamma(k+1)} f^{(k)}(0) + R_{n-1},$$

where

$$R_{n-1} = \int_0^x \frac{f^{(n)}(\tau)(x-\tau)^{n-1}}{(n-1)!} d\tau = \frac{1}{\Gamma(n)} \int_0^x f^{(n)}(\tau)(x-\tau)^{n-1} d\tau = I^n f^{(n)}(x).$$

Now, by using the linearity of the Riemann–Liouville fractional derivative, we obtain

$$D^{\alpha}f(x) = D^{\alpha}\left(\sum_{k=0}^{n-1} \frac{x^k}{\Gamma(k+1)} f^{(k)}(0) + R_{n-1}\right)$$

$$= \sum_{k=0}^{n-1} \frac{D^{\alpha}x^k}{\Gamma(k+1)} f^{(k)}(0) + D^{\alpha}R_{n-1}$$

$$= \sum_{k=0}^{n-1} \frac{\Gamma(k+1)}{\Gamma(k-\alpha+1)} \frac{x^{k-\alpha}}{\Gamma(k+1)} f^{(k)}(0) + D^{\alpha}I^n f^n(x)$$

$$= \sum_{k=0}^{n-1} \frac{x^{k-\alpha}}{\Gamma(k-\alpha+1)} f^{(k)}(0) + I^{n-\alpha} f^n(x)$$

$$= \sum_{k=0}^{n-1} \frac{x^{k-\alpha}}{\Gamma(k-\alpha+1)} f^{(k)}(0) + {}^{C}D^{\alpha}f(x).$$

This means that

$$^{C}D^{\alpha}f(x) = D^{\alpha}f(x) - \sum_{k=0}^{n-1} \frac{x^{k-\alpha}}{\Gamma(k+1-\alpha)} f^{(k)}(0).$$

So the proof is complete.   □

We shall state some properties of the operators I^{α} and $^{C}D^{\alpha}$.

Theorem 2.3.2 *For α, $\beta > 0$ and f as a suitable function, we have*

(i) $I^{\alpha}{}^{C}D^{\alpha}f(t) = f(t) - f(0),\ 0 < \alpha < 1,$

(ii) ${}^{C}D^{\alpha}I^{\alpha}f(t) = f(t),$

(iii) ${}^{C}D^{\alpha}f(t) = I^{1-\alpha}Df(t) = I^{1-\alpha}f'(t),\ 0 < \alpha < 1,$

(iv) ${}^{C}D^{\alpha}{}^{C}D^{\beta}f(t) \neq {}^{C}D^{\alpha+\beta}f(t),$

(v) ${}^{C}D^{\alpha}{}^{C}D^{\beta}f(t) \neq {}^{C}D^{\beta}{}^{C}D^{\alpha}f(t).$

Due to this fact, the concept of sequential fractional derivative is discussed by Miller and Ross [4].

Sequential Fractional Derivative

For $n \in \mathbb{N}$, the sequential fractional derivative for suitable function $f(t)$ is defined by

$$f^{(k\alpha)} := \mathbf{D}^{k\alpha}f(x) = \mathbf{D}^{\alpha}\mathbf{D}^{(k-1)\alpha}f(t),$$

where $k = 1, \cdots, n$, $\mathbf{D}^{0}f(t) = f(t)$, and $\mathbf{D}^{\alpha}$ is any fractional differential operator; here we mention it as ${}^{C}D^{\alpha}$.

Proposition 2.3.3 *When α becomes an integer, the Caputo derivative is a classical derivative.*

Proof We know that

$$^{C}D^{\alpha}f(t) = \frac{1}{\Gamma(n-\alpha)}\int_{a}^{t}(t-s)^{n-\alpha-1}f^{(n)}(s)ds.$$

Taking limit as $\alpha \to n$

$$\lim_{\alpha \to n}{}^{C}D^{\alpha}f(t) = \lim_{\alpha \to n}\frac{1}{\Gamma(n-\alpha)}\int_{a}^{t}(t-s)^{n-\alpha-1}f^{(n)}(s)ds.$$

Taking

$$u = f^{(n)}(s), \quad dv = (t-s)^{n-\alpha-1}ds$$

$$du = f^{(n+1)}(s)ds, \quad v = -\frac{(t-s)^{n-\alpha}}{n-\alpha},$$

we have

$$\lim_{\alpha \to n} {}^C D^\alpha f(t) = \lim_{\alpha \to n} \frac{f^{(n)}(a)(t-a)^{n-\alpha}}{\Gamma(n-\alpha+1)}$$

$$+ \lim_{\alpha \to n} \frac{1}{\Gamma(n-\alpha+1)} \int_a^t (t-s)^{n-\alpha} f^{(n+1)}(s)ds$$

$$= f^{(n)}(a) + \int_a^t f^{(n+1)}(s)ds$$

$$= f^{(n)}(t).$$

$\square$

Proposition 2.3.4 *When α becomes an integer, the Riemann–Liouville fractional derivative is also a classical derivative.*

Proof The relation between Riemann–Liouville and Caputo derivatives is

$$^C D^\alpha f(t) = D^\alpha f(t) - \sum_{k=0}^{n-1} \frac{f^k(a)}{\Gamma(k-\alpha+1)}(t-a)^{k-\alpha}$$

$$D^\alpha f(t) = {}^C D^\alpha f(t) + \sum_{k=0}^{n-1} \frac{f^k(a)}{\Gamma(k-\alpha+1)}(t-a)^{k-\alpha}.$$

Taking limit as $\alpha \to n$

$$\lim_{\alpha \to n} D^\alpha f(t) = \lim_{\alpha \to n} {}^C D^\alpha f(t) + \lim_{\alpha \to n} \sum_{k=0}^{n-1} \frac{f^k(a)}{\Gamma(k-\alpha+1)}(t-a)^{k-\alpha}$$

$$= f^{(n)}(t) + \lim_{\alpha \to n} \frac{f^0(a)}{\Gamma(1-\alpha)}(t-a)^{-\alpha}$$

$$+ \lim_{\alpha \to n} \frac{f^1(a)}{\Gamma(2-\alpha)}(t-a)^{1-\alpha}$$

$$\vdots$$

$$+ \lim_{\alpha \to n} \frac{f^{n-1}(a)}{\Gamma(n-\alpha)}(t-a)^{n-1-\alpha}$$

$$= f^{(n)}(t).$$

Here we used the definition of gamma integral for negative integer. We conclude that α becomes an integer for all fractional derivatives and integral becomes ordinary integer derivative and integral. $\square$

2.4   Examples

From the following example, one can observe that the fractional derivative does not satisfy the commutative property.

Example 2.4.1

$$D^\alpha D^\beta f(t) \neq D^\beta D^\alpha f(t).$$

Let $f(t) = t^{\frac{1}{2}}$, $\alpha = \frac{3}{2}$, $\beta = \frac{1}{2}$. On calculating, we get

$$D^{\frac{1}{2}} t^{\frac{1}{2}} = \frac{\sqrt{\pi}}{2},$$

and

$$D^{\frac{3}{2}} t^{\frac{1}{2}} = D^1 D^{\frac{1}{2}} t^{\frac{1}{2}} = 0.$$

So we get

$$D^{\frac{1}{2}} D^{\frac{3}{2}} t^{\frac{1}{2}} = 0.$$

Now

$$D^{\frac{3}{2}} D^{\frac{1}{2}} t^{\frac{1}{2}} = \frac{\sqrt{\pi}}{2} D^{\frac{3}{2}}(1) = -\frac{t^{-3/2}}{4}.$$

From the above two equations, we observe that

$$D^{\frac{1}{2}} D^{\frac{3}{2}} t^{\frac{1}{2}} \neq D^{\frac{3}{2}} D^{\frac{1}{2}} t^{\frac{1}{2}}.$$

Example 2.4.2 If $n - 1 < \alpha \leq n$ and $\beta > -1$, then

$$I^\alpha t^\beta = \frac{1}{\Gamma(\alpha)} \int_0^t (t - s)^{\alpha-1} s^\beta dt = \frac{t^{\alpha-1}}{\Gamma(\alpha)} \int_0^t \left(1 - \frac{s}{t}\right)^{\alpha-1} s^\beta ds.$$

Further replacing y by s/t, we have

$$I^\alpha t^\beta = \frac{t^{\alpha+\beta}}{\Gamma(\alpha)} \int_0^1 y^\beta (1 - y)^{\alpha-1} dy.$$

Therefore

$$I^\alpha t^\beta = \frac{\Gamma(\beta+1)}{\Gamma(\alpha+\beta+1)} t^{\alpha+\beta}. \tag{2.24}$$

Example 2.4.3 If $n-1 < \alpha \le n$ and $\beta > -1$, then

$$D^\alpha t^\beta = D^n I^{n-\alpha} t^\beta = D^n \left[\frac{\Gamma(\beta+1)}{\Gamma(\beta-\alpha+n+1)} t^{\beta-\alpha+n} \right]$$

$$= \frac{\Gamma(\beta+1)}{\Gamma(\beta-\alpha+n+1)} (\beta-\alpha+n) \cdots (\beta-\alpha-n+1) t^{\beta-\alpha}$$

$$= \frac{\Gamma(\beta+1)}{\Gamma(\beta-\alpha+1)} t^{\beta-\alpha}, \quad \beta > -1.$$

Note that $D^\alpha K = \dfrac{Kt^{-\alpha}}{\Gamma(1-\alpha)}$, where K is a constant. In particular,

$$D^{\frac{1}{2}} t^{\frac{1}{2}} = \frac{\sqrt{\pi}}{2},$$

and the Riemann–Liouville derivative of a constant is not zero. Further $D^1 t^{\frac{1}{2}}$ does not exist, but $D^\alpha t^{\frac{1}{2}}$ exists for $\alpha \le \frac{1}{2}$.

Example 2.4.4 If $n-1 < \alpha \le n$ and $\beta > n-1$, then

$$^C D^\alpha t^\beta = I^{n-\alpha} D^n t^\beta = I^{n-\alpha} \left[\frac{\Gamma(\beta+1)}{\Gamma(\beta-n+1)} t^{\beta-n} \right] = \frac{\Gamma(\beta+1)}{\Gamma(\beta-n+1)} I^{n-\alpha} t^{\beta-n}$$

$$= \frac{\Gamma(\beta+1)}{\Gamma(\beta-n+1)} \times \frac{\Gamma(\beta-n+1)}{\Gamma(\beta-\alpha+1)} t^{\beta-\alpha}$$

$$= \frac{\Gamma(\beta+1)}{\Gamma(\beta-\alpha+1)} t^{\beta-\alpha}, \quad \beta > n-1.$$

As a special case, consider $\alpha = \frac{1}{2}$ and $\beta = 1$; then $f(t) = t$ and

$$^C D^{\frac{1}{2}} t = \frac{1}{\Gamma(\frac{1}{2})} \int_0^t \frac{1}{(t-\tau)^{\frac{1}{2}}} d\tau.$$

Taking into account the properties of the gamma function and using the substitution $u = (t-\tau)^{\frac{1}{2}}$, the final result for the Caputo fractional derivative of the function $f(t) = t$ is obtained as

$$^{C}D^{\frac{1}{2}}t = -\frac{1}{\sqrt{\pi}} \int_0^t \frac{1}{(t-\tau)^{\frac{1}{2}}} d(t-\tau)$$

$$= -\frac{1}{\sqrt{\pi}} \int_{\sqrt{t}}^0 \frac{1}{u} du^2, \quad u = (t-\tau)^{\frac{1}{2}}$$

$$= \frac{1}{\sqrt{\pi}} \int_0^{\sqrt{t}} \frac{2u}{u} du$$

$$= \frac{2}{\sqrt{\pi}}(\sqrt{t} - 0).$$

Thus

$$^{C}D^{\frac{1}{2}}t = \frac{2\sqrt{t}}{\sqrt{\pi}}.$$

Observe that $^{C}D^{\frac{1}{2}} {}^{C}D^{\frac{1}{2}}t = 1$.

Example 2.4.5

$$I^{\alpha}e^{at} = I^{\alpha}\left(\sum_{k=0}^{\infty} \frac{(at)^k}{k!}\right) = \sum_{k=0}^{\infty} \frac{a^k I^{\alpha}t^k}{\Gamma(k+1)}.$$

By using (2.24), we get

$$I^{\alpha}e^{at} = \sum_{k=0}^{\infty} \frac{a^k t^{k+\alpha}}{\Gamma(k+\alpha+1)} = t^{\alpha}E_{1,1+\alpha}(at). \tag{2.25}$$

Example 2.4.6 Let $n-1 < \alpha \le n$. Then

$$D^{\alpha}e^{at} = D^n I^{n-\alpha}e^{at}.$$

By using (2.25), we get

$$D^{\alpha}e^{at} = D^n \left\{ t^{n-\alpha}E_{1,1+n-\alpha}(at) \right\} = \sum_{k=0}^{\infty} \frac{a^k D^n t^{k+n-\alpha}}{\Gamma(k+n-\alpha+1)}$$

$$= t^{-\alpha}E_{1,1-\alpha}(at).$$

Example 2.4.7 Let $n-1 < \alpha \le n$. Then

$$^{C}D^{\alpha}e^{at} = I^{n-\alpha}D^n e^{at} = a^n I^{n-\alpha}e^{at}.$$

By using (2.25), we get

$$^{C}D^{\alpha}e^{at} = a^n t^{n-\alpha}E_{1,1+n-\alpha}(at).$$

Example 2.4.8

$$I^\alpha \sin at = I^\alpha \left(\sum_{k=0}^{\infty} \frac{(-1)^k (at)^{2k+1}}{\Gamma(2k+2)} \right) = \sum_{k=0}^{\infty} \frac{(-1)^k a^{2k+1}}{\Gamma(2k+2)} I^\alpha t^{2k+1}.$$

By using (2.25), we get

$$I^\alpha \sin at = \sum_{k=0}^{\infty} \frac{(-1)^k a^{2k+1} t^{2k+\alpha+1}}{\Gamma(2k+\alpha+2)} = at^{\alpha+1} E_{2,2+\alpha}(-a^2 t^2). \qquad (2.26)$$

Example 2.4.9 Let $n - 1 < \alpha \le n$.

$$D^\alpha \sin at = D^n I^{n-\alpha} \sin at,$$

By using (2.26), we get

$$\begin{aligned}
D^\alpha \sin at &= D^n \left\{ at^{n-\alpha+1} E_{2,2+n-\alpha}(-a^2 t^2) \right\} \\
&= \sum_{k=0}^{\infty} \frac{(-1)^k a^{2k+1}}{\Gamma(2k+n-\alpha+2)} D^n t^{2k+n-\alpha+1} \\
&= \sum_{k=0}^{\infty} \frac{(-1)^k a^{2k+1} t^{2k+1-\alpha}}{\Gamma(2k+2-\alpha)} \\
&= at^{1-\alpha} E_{2,2-\alpha}(-a^2 t^2).
\end{aligned}$$

Example 2.4.10 Let $n - 1 < \alpha \le n$. Then

$${}^C D^\alpha \sin at = I^{n-\alpha} D^n \sin at = \sum_{k=0}^{\infty} \frac{(-1)^k a^{2k+1}}{\Gamma(2k-n+2)} I^{n-\alpha} t^{2k-n+1}.$$

By using (2.26), we get

$${}^C D^\alpha \sin at = \sum_{k=0}^{\infty} \frac{(-1)^k a^{2k+1} t^{2k+1-\alpha}}{\Gamma(2k+2-\alpha)} = at^{1-\alpha} E_{2,2-\alpha}(-a^2 t^2).$$

Next we show that the product rule for integer order derivative is not true for the fractional derivative.

Example 2.4.11 Consider

$$D^\alpha (e^{at} e^{bt}) = D^\alpha (e^{(a+b)t}).$$

Using the definition, we have

$$D^\alpha(e^{(a+b)t}) = (a+b)^\alpha e^{(a+b)t}$$

but if you use $D^\alpha(f(t)g(t)) = D^\alpha f(t)g(t) + f(t)D^\alpha g(t)$, we get

$$D^\alpha(e^{at}e^{bt}) = D^\alpha e^{at}.e^{bt} + e^{at}D^\alpha e^{bt}$$
$$= (a^\alpha + b^\alpha)e^{(a+b)t} \neq (a+b)^\alpha e^{(a+b)t}.$$

Therefore

$$D^\alpha(f(t)g(t)) \neq D^\alpha f(t)g(t) + f(t)D^\alpha g(t).$$

Similarly the chain rule for integer order derivative is not true for the fractional derivative.

$$\frac{d^\alpha f(g(t))}{dt^\alpha} \neq \frac{d^\alpha f(g(t))}{dg^\alpha}\frac{d^\alpha g(t)}{dt^\alpha}.$$

Remark: [5] In 1914, Ramanujan introduced the idea of fractional differentiation of x^n by adopting the notation D for ordinary differentiation and using the gamma function. He obtained the following identity

$$D^m x^n = \frac{\Gamma(n+1)}{\Gamma(n-m+1)}x^{n-m}$$

and for $m = \frac{1}{2}$,

$$D^{\frac{1}{2}}x^n = \frac{\Gamma(n+1)}{\Gamma(n+\frac{1}{2})}x^{n-\frac{1}{2}}.$$

Example 2.4.12   (R-L fractional derivative of a basic power function)

Let us take $f(t) = t^k$. Then the first derivative is

$$f'(t) = \frac{d}{dt}f(t) = kt^{k-1}.$$

Repeating this gives the more general result that

$$\frac{d^\alpha}{dt^\alpha}t^k = \frac{k!}{(k-\alpha)!}t^{k-\alpha}$$

which, after replacing the factorials with the gamma function, leads us to

$$\frac{d^\alpha}{dt^\alpha}t^k = \frac{\Gamma(k+1)}{\Gamma(k-\alpha+1)}t^{k-\alpha}, \quad k \geq 0.$$

For $k = 1$ and $\alpha = \frac{1}{2}$, we obtain the half-derivative of the function t as

$$\frac{d^{\frac{1}{2}}}{dt^{\frac{1}{2}}}t = \frac{\Gamma(1+1)}{\Gamma(1-\frac{1}{2}+1)}t^{1-\frac{1}{2}} = \frac{\Gamma(2)}{\Gamma(\frac{3}{2})}t^{\frac{1}{2}} = \frac{1}{\frac{\sqrt{\pi}}{2}}t^{\frac{1}{2}}.$$

Repeating this process yields

$$\frac{d^{\frac{1}{2}}}{dt^{\frac{1}{2}}}\frac{2t^{\frac{1}{2}}}{\sqrt{\pi}} = \frac{2}{\sqrt{\pi}}\frac{\Gamma(1+\frac{1}{2})}{\Gamma(\frac{1}{2}-\frac{1}{2}+1)}t^{\frac{1}{2}-\frac{1}{2}} = \frac{2}{\sqrt{\pi}}\frac{\Gamma(\frac{3}{2})}{\Gamma(1)}t^0 = \frac{2\frac{\sqrt{\pi}}{2}}{\sqrt{\pi}}t^0 = 1,$$

$$\left(\text{because } \Gamma(\frac{3}{2}) = \frac{1}{2}\sqrt{\pi} \text{ and } \Gamma(1) = 1\right)$$

which is indeed the expected result of

$$\left(\frac{d^{\frac{1}{2}}}{dt^{\frac{1}{2}}}\frac{d^{\frac{1}{2}}}{dt^{\frac{1}{2}}}\right)t = \frac{d}{dt}t = 1.$$

For negative integer power k, the gamma function is undefined and we have to use the following relation:

$$\frac{d^\alpha}{dt^\alpha}t^{-k} = (-1)^\alpha\frac{\Gamma(k+\alpha)}{\Gamma(k)}t^{-(k+\alpha)} \text{ and } k \geq 0.$$

This extension of the above differential operator need not be constrained only to real powers. For example, the $(1+i)$th derivative of the $(1-i)$th derivative yields the second derivative. Also notice that setting negative values for α yields integrals. For a general function f(t) and $0 < \alpha < 1$, the complete fractional derivative is

$$D^\alpha f(t) = \frac{1}{\Gamma(1-\alpha)}\frac{d}{dt}\int_0^t \frac{f(s)}{(t-s)^\alpha}ds.$$

For arbitrary α, since the gamma function is undefined for arguments whose real part is a negative integer and whose imaginary part is zero, it is necessary to apply the fractional derivative after the integer derivative has been performed. For example,

$$D^{\frac{3}{2}}f(t) = D^{\frac{1}{2}}D^1 f(t) = D^{\frac{1}{2}}\frac{d}{dt}f(t).$$

Observe that, for zero initial conditions, the two derivatives are the same. It is known that Riemann–Liouville derivative leads to difficulties while applying to real-world problems, especially in the context of initial conditions. Hence relatively recent fractional derivative introduced by Caputo is extensively used in applications as the initial conditions involved are physically meaningful. The difference between the Riemann–Liouville definition and the Caputo definition is that the Caputo derivative of a constant K is 0, that is, $^C D^\alpha K = 0$, whereas the Riemann–Liouville fractional derivative of a constant K need not be equal to zero, that is, $D^\alpha K = \frac{Kt^{-\alpha}}{\Gamma(1-\alpha)}$.

Example 2.4.13 In the fractional calculus, the law of exponents is known to be generally true for the operators of fractional integration due to their semigroup property. In general, both the operators of fractional differentiation, D^α and $^C D^\alpha$, do not satisfy either the semigroup property or the (weaker) commutative property. To show how the law of exponents does not necessarily hold for the standard fractional derivative, we provide two simple examples (with power functions) for which

(a) $D^\alpha D^\beta f(t) = D^\beta D^\alpha f(t) \neq D^{\alpha+\beta} f(t)$,
(b) $D^\alpha D^\beta g(t) \neq D^\beta D^\alpha g(t) \neq D^{\alpha+\beta} g(t)$.

For (a), let us take $f(t) = t^{-\frac{1}{2}}$ and $\alpha = \beta = \frac{1}{2}$. Then, using the property, we get $D^{\frac{1}{2}} f(t) = 0$, $D^{\frac{1}{2}} D^{\frac{1}{2}} f(t) = 0$, but $D^{\frac{1}{2}+\frac{1}{2}} f(t) = Df(t) = -\frac{t^{-\frac{3}{2}}}{2}$. For (b), let us take $g(t) = t^{\frac{1}{2}}$ and $\alpha = \frac{1}{2}, \beta = \frac{3}{2}$. Then, from Example (2.4.1), we get $D^{\frac{1}{2}} g(t) = \frac{\sqrt{\pi}}{2}$, $D^{\frac{3}{2}} g(t) = 0$, but $D^{\frac{1}{2}} D^{\frac{3}{2}} g(t) = 0$, $D^{\frac{3}{2}} D^{\frac{1}{2}} g(t) = -\frac{t^{-\frac{3}{2}}}{4}$ and $D^{\frac{1}{2}+\frac{3}{2}} g(t) = D^2 g(t) = -\frac{t^{-\frac{3}{2}}}{4}$.

2.5   Exercises

2.1. Show that, for $\alpha > 0$, $I^\alpha(xf(x)) = xI^\alpha f(x) - \alpha I^{\alpha+1} f(x)$.
2.2. Find the R-L derivative of order $\frac{1}{2}$ of the function $f(x) = \ln x$.
2.3. Show that $D^\alpha(\cos ax) = a^\alpha \cos(ax + \alpha\pi/2)$ for $a, \alpha \in \mathbb{R}$.
2.4. Show that $\sqrt{2} D^{\frac{1}{2}} \sin x = \sin x + \cos x$.
2.5. Calculate the Caputo derivative of order α with $n - 1 < \alpha \leq n$ for the following functions: (i) $f(x) = (1 + x)^m$, (ii) $f(x) = e^x$.
2.6. Evaluate $I^\alpha f(x) = 1 + f(x)$ for $\alpha > 0$ and $f(x)$ is a continuous function.
2.7. Let $f(t)$ be a continuous function and $0 < \alpha \leq 1$. Solve the integral equation $\int_0^t (t - s)^{-\alpha} f(s) ds = 1$.
2.8. Let $a, b \in \mathbb{R}, 0 < \alpha \leq 1$. Show that the integral equation

$$x(t) = \frac{at^{1-\alpha}}{\Gamma(2 - \alpha)} - \frac{b}{\Gamma(1 - \alpha)} \int_0^t (t - \tau)^{-\alpha} x(\tau) d\tau$$

has a solution $x(t) = \frac{a}{b}[1 - E_{1-\alpha}(-bt^{1-\alpha})]$.
2.9. Show that the solution of the integrodifferential equation

$$\int_0^t (t - \tau)^{-\frac{1}{2}} f'(\tau) d\tau = t$$

is $f(t) = \frac{4}{3\pi} t \sqrt{t}$.

2.10. Let $f(t)$ be a continuous function such that $f'(0) = 0$ and $0 < \alpha \le 1$. Show that the solution of the integrodifferential equation $^C D^\alpha f(t) = 1 + f(t)$ is given by $f(t) = t^\alpha E_{\alpha, \alpha+1}(t^\alpha)$.

References

1. Kilbas, A.A., Srivastava, H.M., Trujillo, J.J.: Theory and Applications of Fractional Differential Equations. Elsevier, Amsterdam (2006)
2. Samko, S.G., Kilbas, A.A., Marichev, O.I.: Fractional Integrals and Derivatives. Theory and Applications. Gordon and Breach Science Publishers, Amsterdam (1993)
3. Caputo, M.: Linear model of dissipation whose Q is almost frequency independent-II. Geophys. J. Roy. Astron. Soc. **13**, 529–539 (1967)
4. Miller, K., Ross, B.: An Introduction to the Fractional Calculus and Fractional Differential Equations. John Wiley and Sons Inc, New York (1993)
5. Ranganathan, S.R.: Ramanujan. The Man and The Mathematician. Ess-Ess Publications, New Delhi (1967)

Chapter 3
Fractional Differential Equations

Abstract Fractional differential equations appear more frequently in different areas of science and engineering. As motivation, the occurrence of these equations in different fields of study is indicated. Solution representations for linear fractional differential equations are obtained with the help of the Mittag–Leffler matrix functions. Existence of solutions for nonlinear fractional differential equations and nonlinear fractional damped equations are established by using fixed point theorems. Several examples are provided and a set of exercises is given.

Keywords Mittag–Leffler matrix functions · Linear equations · Solution representation · Existence of solutions · Nonlinear fractional differential equations · Fixed point method

3.1 Motivation

Fractional differential equations appear more frequently in different areas of science and engineering. In fact, real-world processes generally or most likely result in fractional order systems. The main reason for using the integer order models was the absence of solution methods for fractional differential equations. The most important advantage of using fractional differential equation is their nonlocal property. It is well known that the integer order differential operator is a local operator but the fractional order differential operator is nonlocal. This means that the next state of a system depends not only upon its current state but also upon all its past states. Many real-world systems are better characterized by using a non-integer order dynamic model based on fractional calculus. Fractional calculus can be an aid for explanation of discontinuity formation and singularity formation is an enriching thought experiment. We may say that nature works with fractional derivatives. In fact, fractional differential equations are alternative models of nonlinear differential equations [1]. More interesting facts and the importance of fractional differential equations can be found in [2–9]. In the following models, the fractional derivatives are taken in the Caputo sense.

(i) Fractional Order HIV/AIDS Model

The integer order HIV/AIDS model is recently reconstructed as the fractional order model [10]. In that, they first divided the total population into a susceptible class of size S and an infectious class before the onset of AIDS and a full-blown AIDS group of size A which is removed from the active population. Based on the assumptions that the infectious period is very long (≥ 10 years), we further consider several stages of the infectious period. For simplicity, we consider two stages according to clinic stages, that is, the asymptomatic phase (I) and the symptomatic phase (J). Thus, the model can be described by

$$\frac{d^\alpha S}{dt^\alpha} = \mu K - c\beta(I + bJ)S - \mu S,$$

$$\frac{d^\alpha I}{dt^\alpha} = c\beta(I + bJ)S - (\mu + k_1)I + \gamma J,$$

$$\frac{d^\alpha J}{dt^\alpha} = k_1 I - (\mu + k_2 + \gamma)J,$$

$$\frac{d^\alpha A}{dt^\alpha} = k_2 J - (\mu + d)A,$$

$$S(\delta) = S_0, \quad I(\delta) = I_0, \quad J(\delta) = J_0, \quad A(\delta) = A_0,$$

where μK is the recruitment rate of the population; μ is the number of death rate constant; c is the average number of constants of an individual per unit time; β and $b\beta$ are probabilities of disease transmission per contact by an infective in the first stage and in the second stage, respectively; k_1 and k_2 are the transfer rate constants from the asymptomatic phase I to the symptomatic phase J and from the symptomatic phase to the AIDS case, respectively; γ is treatment rate from the symptomatic phase J to the asymptomatic phase I; d is the disease-related death rate of the AIDS cases.

(ii) Fractional Model of Tumor–Immune System

Immune system is one of the most fascinating schemes from the point of view of biology and mathematics. The immune system is complex, intricate, and interesting. It is known to be multifunctional and multipathway, so most immune effectors do more than one job. Also, each function of the immune system is typically done by more than one effector which makes it more robust. Studying immune system cancer interactions is an important topic. The reason for using fractional order differential equations is that they are naturally related to systems with memory which exist in tumor–immune interactions.

The model includes two immune effectors: $E_1(t)$, $E_2(t)$ (such as cytotoxic T cells and natural killer cells) interacting with the cancer cells, $T(t)$, with a Holling function of type III. (Holling type III describes the response of predators to prey depressed at low prey density, then levels off with a further increase in prey density.)

The model takes the form [11]

$$\frac{d^\alpha T}{dt^\alpha} = aT - r_1 T E_1 - r_2 T E_2,$$

$$\frac{d^\alpha E_1}{dt^\alpha} = -d_1 E_1 + \frac{T^2 E_1}{T^2 + k_1},$$

$$\frac{d^\alpha E_2}{dt^\alpha} = -d_2 E_2 + \frac{T^2 E_2}{T^2 + k_2},$$

where $0 < \alpha \leq 1$ and $a, r_1, r_2, d_1, d_2, k_1,$ and k_2 are positive constants. The interaction terms in the second and third equations of the above model satisfy the cross-reactivity property of the immune system.

(iii) Fractional Model of Electrical Circuits

Consider linear electrical circuits composed of resistors, supercondensators (ultracapacitors), coils, and voltage (current) sources. As the state varies, the voltage across the supercondensator, currents in the coils are usually chosen. It is well known that the current $i(t)$ in supercondensator is related with its voltage $u_C(t)$ by the formula

$$i_C(t) = C \frac{d^\alpha u_C(t)}{dt^\alpha}, \quad 0 < \alpha < 1,$$

where C is the capacity of the supercondensator. Similarly, the voltage $u_L(t)$ on the coil is related with its current $i_L(t)$ by the formula

$$u_L(t) = L \frac{d^\beta i_L(t)}{dt^\beta}, \quad 0 < \beta < 1,$$

where L is the inductance of the coil. The advantages of fractional derivatives become apparent in modeling electrical properties of real materials [12].

(iv) Fractional Oscillator with Damping

The fractional derivative version of the conventional single degree of freedom oscillator is referred to as the single degree of freedom fractional oscillator. The equation of motion is given by

$$m\ddot{\vartheta}(t) + c^C D^\alpha \vartheta(t) + k\vartheta(t) = f(t),$$

where m is the mass, c the damping coefficient, k the stiffness, ϑ the displacement, and f the forcing function. Further Bagley and Torvik (see [13]) analyzed the viscoelastically damped structures by means of fractional calculus and introduced the Bagley–Torvik equation of order $\alpha = 1/2$ and $\alpha = 3/2$ to study the motion of a rigid plate in a Newtonian fluid. The generic form of Bayley–Torvik equation is given by

$$aD^2y(t) + b^C D^\alpha y(t) + cy(t) = f(t), \ t \in [0, T],$$
$$D^k y(0) = c_k, \ k = 0, 1,$$

where $\alpha = 1/2$ or $\alpha = 3/2$, $a > 0$ and $b, c \in \mathbb{R}$.

(v) Fractional Electrochemistry Model

In electrochemistry, the idea of a half-order fractional integral of the current field was known. Oldham and Spanier [14] suggested the replacement of the classical integer order Fick's law describing the diffusion of electroactive species toward the electrodes by a fractional order integral law in the form

$$D^{-1/2}i(t) = kc_0 \left[\left\{ 1 - \frac{C(0, t)}{c_0} \right\} + \frac{\sqrt{k}}{R} D^{-1/2} \left\{ 1 - \frac{C(0, t)}{c_0} \right\} \right],$$

where c_0 is the uniform concentration of electro-active species, k is the diffusion coefficient and k and R are constants.

3.2 Equation with Constant Coefficient

Consider the fractional differential equation of the form

$$^C D^\alpha x(t) = ax(t) + f(t), \ 0 < t \le T, \ a \in \mathbb{R}, \tag{3.1}$$
$$x(0) = x_0,$$

where $0 < \alpha < 1$ and $f(t)$ is a continuous function on $[0, T]$. Taking I^α on both sides of (3.1), we get the corresponding Volterra integral equation

$$x(t) = x_0 + \frac{a}{\Gamma(\alpha)} \int_0^t (t - s)^{\alpha-1} x(s) \mathrm{d}s + \frac{1}{\Gamma(\alpha)} \int_0^t (t - s)^{\alpha-1} f(s) \mathrm{d}s.$$

Now we apply the method of successive approximations to solve this integral equation. For that, we set

$$x_0(t) = x_0,$$
$$x_m(t) = x_0 + \frac{a}{\Gamma(\alpha)} \int_0^t (t - s)^{\alpha-1} x_{m-1}(s) \mathrm{d}s + \frac{1}{\Gamma(\alpha)} \int_0^t (t - s)^{\alpha-1} f(s) \mathrm{d}s,$$

for $m = 1, 2, \dots$.

We find for $x_1(t)$, that is,

$$x_1(t) = x_0 + \frac{a}{\Gamma(\alpha)}\int_0^t (t-s)^{\alpha-1}x_0 ds + \frac{1}{\Gamma(\alpha)}\int_0^t (t-s)^{\alpha-1}f(s)ds$$

$$= x_0 + \frac{at^\alpha}{\Gamma(\alpha+1)}x_0 + \frac{1}{\Gamma(\alpha)}\int_0^t (t-s)^{\alpha-1}f(s)ds$$

$$= \sum_{k=0}^{1}\frac{a^k t^{\alpha k}}{\Gamma(\alpha k+1)}x_0 + \frac{1}{\Gamma(\alpha)}\int_0^t (t-s)^{\alpha-1}f(s)ds.$$

Similarly, we find for $x_2(t)$,

$$x_2(t) = x_0 + \frac{a}{\Gamma(\alpha)}\int_0^t (t-s)^{\alpha-1}x_1(s)ds + \frac{1}{\Gamma(\alpha)}\int_0^t (t-s)^{\alpha-1}f(s)ds$$

$$= x_0 + \sum_{k=0}^{1}\frac{a^k+1}{\Gamma(\alpha)\Gamma(\alpha k+1)}\int_0^t (t-s)^{\alpha-1}s^{\alpha k}x_0 ds$$

$$+ \frac{a}{\Gamma(\alpha)}\int_0^t (t-s)^{\alpha-1}\left(\frac{1}{\Gamma(\alpha)}\int_0^s (s-\tau)^{\alpha-1}f(\tau)d\tau\right)ds$$

$$+ \frac{1}{\Gamma(\alpha)}\int_0^t (t-s)^{\alpha-1}f(s)ds$$

$$= \sum_{k=0}^{2}\frac{a^k t^{\alpha k}}{\Gamma(\alpha k+1)}x_0 + \int_0^t \sum_{k=1}^{2}\frac{a^{k-1}(t-s)^{\alpha k-1}}{\Gamma(\alpha k)}f(s)ds$$

$$= \sum_{k=0}^{2}\frac{a^k t^{\alpha k}}{\Gamma(\alpha k+1)}x_0 + \int_0^t \sum_{k=1}^{2}\frac{a^{k-1}(t-s)^{\alpha k-1}}{\Gamma(\alpha k)}f(s)ds.$$

Continuing this process, we derive the following relation for $x_m(t)$, $m \in \mathbb{N}$, as

$$x_m(t) = \sum_{k=0}^{m}\frac{a^k t^{\alpha k}}{\Gamma(\alpha k+1)}x_0 + \int_0^t \sum_{k=1}^{m-1}\frac{a^{k-1}(t-s)^{\alpha k-1}}{\Gamma(\alpha k)}f(s)ds$$

$$= \sum_{k=0}^{m}\frac{a^k t^{\alpha k}}{\Gamma(\alpha k+1)}x_0 + \int_0^t \sum_{k=0}^{m-1}\frac{a^k(t-s)^{\alpha k+\alpha-1}}{\Gamma(\alpha k+\alpha)}f(s)ds.$$

Taking the limit as $m \to \infty$, we obtain the explicit form of $x(t)$ as

$$x(t) = \sum_{k=0}^{\infty}\frac{a^k t^{\alpha k}}{\Gamma(\alpha k+1)}x_0 + \int_0^t (t-s)^{\alpha-1}\sum_{k=0}^{\infty}\frac{a^k(t-s)^{\alpha k}}{\Gamma(\alpha k+\alpha)}f(s)ds.$$

We write the above equation in terms of the Mittag–Leffler function and the solution of (3.1) is

$$x(t) = E_\alpha(at^\alpha)x_0 + \int_0^t (t-s)^{\alpha-1} E_{\alpha,\alpha}(a(t-s)^\alpha)f(s)ds. \qquad (3.2)$$

If the inhomogeneous term $f(t) = 0$, then the Eq. (3.1) is called homogeneous equation of the form

$$^C D^\alpha x(t) = ax(t), \ 0 < t \le T, \ a \in \mathbb{R},$$
$$x(0) = x_0,$$

and the solution is

$$x(t) = E_\alpha(at^\alpha)x_0.$$

The solution of the Cauchy problem

$$^C D^\alpha x(t) - ax(t) = f(t), \quad x(0) = x_0, \quad x'(0) = y_0,$$

where $1 < \alpha < 2$ and $x_0, y_0, a \in \mathbb{R}$ is of the form

$$x(t) = E_\alpha(at^\alpha)x_0 + tE_{\alpha,2}(at^\alpha)y_0 + \int_0^t (t-s)^{\alpha-1} E_{\alpha,\alpha}(a(t-s)^\alpha)f(s)ds.$$

In particular, the solution to the equation

$$^C D^\alpha x(t) - ax(t) = 0, \quad x(0) = x_0, \quad x'(0) = y_0,$$

is given by

$$x(t) = E_\alpha(at^\alpha)x_0 + tE_{\alpha,2}(at^\alpha)y_0.$$

3.3 Equation with Matrix Coefficient

If the constant a is replaced by an $n \times n$ matrix A and $x(t)$, $f(t)$ are n vectors, then the Eq. (3.1) becomes vector fractional differential equation

$$^C D^\alpha x(t) = Ax(t) + f(t), \ 0 < t \le T, \qquad (3.3)$$
$$x(0) = x_0,$$

and the solution takes the following form:

$$x(t) = E_\alpha(At^\alpha)x_0 + \int_0^t (t-s)^{\alpha-1} E_{\alpha,\alpha}(A(t-s)^\alpha)f(s)ds. \qquad (3.4)$$

Similarly, when $f = 0$ in (3.3), the solution (3.4) takes the form

$$x(t) = E_\alpha(At^\alpha)x_0.$$

Observe that

$$^C D^\alpha x(t) = {}^C D^\alpha E_\alpha(At^\alpha)x_0 = A E_\alpha(At^\alpha)x_0 = Ax(t).$$

Next, we obtain the solution representation by using the Laplace transform technique.

Consider the linear fractional differential equation of the form

$$^C D^\alpha x(t) = Ax(t) + f(t), \quad t \in J = [0, T], \tag{3.5}$$
$$x(0) = x_0,$$

where $0 < \alpha < 1, x \in \mathbb{R}^n$, A is an $n \times n$ matrix and $f(t)$ is continuous on J. Applying Laplace transform on both sides and using the Laplace transform of Caputo derivative, we get

$$\lambda^\alpha X(\lambda) - \lambda^{\alpha-1}x(0) = AX(\lambda) + F(\lambda),$$
$$X(\lambda) = \frac{\lambda^{\alpha-1}x(0)}{[\lambda^\alpha I - A]} + \frac{F(\lambda)}{[\lambda^\alpha I - A]}.$$

Applying inverse Laplace transform on both sides, we have

$$\mathcal{L}^{-1}\{X(\lambda)\}(t) = \mathcal{L}^{-1}\left\{\lambda^{\alpha-1}(\lambda^\alpha I - A)^{-1}\right\}(t)x_0$$
$$+\mathcal{L}^{-1}\left\{\frac{F(\lambda)}{[\lambda^\alpha I - A]}\right\}(t).$$

Inserting Laplace transform of Mittag–Leffler function, we get the solution as

$$x(t) = E_\alpha(At^\alpha)x_0 + \left[\frac{d}{dt}\int_0^t \frac{(t-s)^{\alpha-1}}{\Gamma(\alpha)} E_\alpha(As^\alpha)ds\right] * f(t)$$
$$= E_\alpha(At^\alpha)x_0 + \int_0^t s^{\alpha-1} E_{\alpha,\alpha}(As^\alpha)f(t-s)ds$$
$$= E_\alpha(At^\alpha)x_0 + \int_0^t (t-s)^{\alpha-1} E_{\alpha,\alpha}(A(t-s)^\alpha)f(s)ds.$$

Therefore, the solution is given by

$$x(t) = E_\alpha(At^\alpha)x_0 + \int_0^t (t-s)^{\alpha-1} E_{\alpha,\alpha}(A(t-s)^\alpha)f(s)ds. \tag{3.6}$$

Solution Verification:

Now we verify that the solution satisfies the fractional differential equation (3.5). We know that

$$^C D^\alpha E_\alpha(At^\alpha) = A E_\alpha(At^\alpha),$$

$$\frac{d}{dt} t^\alpha E_{\alpha,\alpha+1}(At^\alpha) = \sum_{k=0}^{\infty} \frac{A^k}{\Gamma(\alpha k + \alpha + 1)} \frac{d}{dt} t^{\alpha k + \alpha}$$

$$= \sum_{k=0}^{\infty} \frac{A^k(\alpha k + \alpha)t^{\alpha k + \alpha - 1}}{\Gamma(\alpha k + \alpha + 1)},$$

$$\frac{d}{dt} t^\alpha E_{\alpha,\alpha+1}(At^\alpha) = t^{\alpha-1} E_{\alpha,\alpha}(At^\alpha)$$

$$\frac{d}{dt} E_\alpha(At^\alpha) = \sum_{k=0}^{\infty} \frac{A^k}{\Gamma(\alpha k + 1)} \frac{d}{dt} t^{\alpha k}$$

$$= \sum_{k=0}^{\infty} \frac{A^k}{\Gamma(\alpha k + 1)} \alpha k t^{\alpha k - 1}$$

$$= \sum_{k=0}^{\infty} \frac{A^k}{\Gamma(\alpha k)} t^{\alpha k - 1}$$

$$= \sum_{k=1}^{\infty} \frac{A^{k+1}}{\Gamma(\alpha k + \alpha)} t^{\alpha k + \alpha - 1}$$

$$= A t^{\alpha-1} E_{\alpha,\alpha}(At^\alpha),$$

$$\int_0^t (t-s)^{\alpha-1} E_{\alpha,\alpha}(A(t-s)^\alpha) ds = \sum_{k=0}^{\infty} \frac{A^k}{\Gamma(\alpha k + \alpha)} \int_0^t (t-s)^{\alpha k + \alpha - 1} ds$$

$$= \sum_{k=0}^{\infty} \frac{A^k}{\Gamma(\alpha k + \alpha)} \cdot \frac{(t-s)^{\alpha k + \alpha}}{\alpha k + \alpha}$$

$$= (t-s)^\alpha E_{\alpha,\alpha+1}(A(t-s)^\alpha).$$

Taking $^C D^\alpha$ on both sides of (3.6),

$$^C D^\alpha x(t) = {}^C D^\alpha E_\alpha(At^\alpha) x_0 \qquad\qquad (3.7)$$

$$+ {}^C D^\alpha \int_0^t (t-s)^{\alpha-1} E_{\alpha,\alpha}(A(t-s)^\alpha) f(s) ds$$

$$^{C}D^{\alpha}x(t) = L_1 + L_2$$

$$\text{where } L_1 = {}^{C}D^{\alpha}E_{\alpha}(At^{\alpha})x_0,$$

$$= AE_{\alpha}(At^{\alpha})x_0,$$

$$\text{and } L_2 = {}^{C}D^{\alpha}\int_0^t (t-s)^{\alpha-1}E_{\alpha,\alpha}(A(t-s)^{\alpha})f(s)ds,$$

$$= \frac{1}{\Gamma(1-\alpha)}\int_0^t (t-s)^{-\alpha}\left(\frac{d}{ds}\right.$$

$$\left.\int_0^s (s-\tau)^{\alpha-1}E_{\alpha,\alpha}(A(s-\tau)^{\alpha})f(\tau)d\tau\right)ds. \qquad (3.8)$$

Consider

$$\frac{d}{ds}\int_0^s (s-\tau)^{\alpha-1}E_{\alpha,\alpha}(A(s-\tau)^{\alpha})f(\tau)d\tau$$

and take

$$u = f(\tau), \quad dv = (s-\tau)^{\alpha-1}E_{\alpha,\alpha}(A(s-\tau)^{\alpha})d\tau$$

$$du = f'(\tau)d\tau, \quad v = -(s-\tau)^{\alpha-1}E_{\alpha,\alpha+1}(A(s-\tau)^{\alpha}).$$

Then

$$\frac{d}{ds}\int_0^s (s-\tau)^{\alpha-1}E_{\alpha,\alpha}(A(s-\tau)^{\alpha})f(\tau)d\tau$$

$$= \frac{d}{ds}\left[-(s-\tau)^{\alpha-1}E_{\alpha,\alpha+1}(A(s-\tau)^{\alpha})f(\tau)\Big|_0^s\right.$$

$$\left.+\int_0^s (s-\tau)^{\alpha}E_{\alpha,\alpha+1}(A(s-\tau)^{\alpha})f'(\tau)d\tau\right]$$

$$= \frac{d}{ds}\left[-0.f(s) + s^{\alpha}E_{\alpha,\alpha+1}(A(s-0)^{\alpha})f(0)\right.$$

$$\left.+\int_0^s (s-\tau)^{\alpha}E_{\alpha,\alpha+1}(A(s-\tau)^{\alpha})f'(\tau)d\tau\right]$$

$$= \frac{d}{ds}\left[s^{\alpha}E_{\alpha,\alpha+1}(As^{\alpha})f(0)\right]$$

$$+\frac{d}{ds}\int_0^s (s-\tau)^{\alpha}E_{\alpha,\alpha+1}(A(s-\tau)^{\alpha})f'(\tau)d\tau$$

$$= s^{\alpha-1}E_{\alpha,\alpha}(As^{\alpha})f(0)$$

$$+\int_0^s (s-\tau)^{\alpha-1}E_{\alpha,\alpha}(A(s-\tau)^{\alpha})f'(\tau)d\tau.$$

Thus,

$$\frac{d}{ds}\int_0^s (s-\tau)^{\alpha-1}\ E_{\alpha,\alpha}(A(s-\tau)^\alpha)f(\tau)d\tau = s^{\alpha-1}E_{\alpha,\alpha}(As^\alpha)f(0)$$

$$+\int_0^s (s-\tau)^{\alpha-1}E_{\alpha,\alpha}(A(s-\tau)^\alpha)f'(\tau)d\tau$$

and from (3.8)

$$L_2 = \frac{1}{\Gamma(1-\alpha)}\int_0^t (t-s)^{-\alpha}s^{\alpha-1}E_{\alpha,\alpha}(As^\alpha)f(0)ds$$

$$+\frac{1}{\Gamma(1-\alpha)}\int_0^t (t-s)^{-\alpha}ds\int_0^s (s-\tau)^{\alpha-1}E_{\alpha,\alpha}(As^\alpha)f'(\tau)d\tau. \quad (3.9)$$

Let

$$L_2 = I_1 + I_2,$$

where

$$I_1 = \frac{1}{\Gamma(1-\alpha)}\int_0^t (t-s)^{-\alpha}s^{\alpha-1}E_{\alpha,\alpha}(As^\alpha)f(0)ds$$

$$= \frac{1}{\Gamma(1-\alpha)}\int_0^t (t-s)^{-\alpha}s^{\alpha-1}\sum_{k=0}^\infty \frac{A^k s^{\alpha k}}{\Gamma(\alpha k+\alpha)}f(0)ds$$

$$= \frac{f(0)}{\Gamma(1-\alpha)}\sum_{k=0}^\infty \frac{A^k}{\Gamma(\alpha k+\alpha)}\int_0^t (t-s)^{-\alpha}s^{\alpha k+\alpha-1}ds$$

Taking $dy = -\frac{ds}{t}, y = \frac{t-s}{t}, 1-y = \frac{s}{t}, s : 0 \to t, y : 1 \to 0,$

$$I_1 = \frac{f(0)}{\Gamma(1-\alpha)}\sum_{k=0}^\infty \frac{A^k}{\Gamma(\alpha k+\alpha)}\int_0^1 (yt)^{-\alpha}((1-y)t)^{\alpha k+\alpha-1}tdy$$

$$= \frac{f(0)}{\Gamma(1-\alpha)}\sum_{k=0}^\infty \frac{A^k}{\Gamma(\alpha k+\alpha)}t^{\alpha k}\int_0^1 y^{-\alpha+-1}(1-y)^{\alpha k+\alpha-1}dy$$

$$= f(0)E_{\alpha,1}(At^\alpha)$$

$$= f(0)E_\alpha(At^\alpha);$$

and

$$I_2 = \frac{1}{\Gamma(1-\alpha)}\int_0^t f'(\tau)d\tau\sum_{k=0}^\infty \frac{A^k}{\Gamma(\alpha k+\alpha)}\int_\tau^t (t-s)^{-\alpha}(s-\tau)^{\alpha k+\alpha-1}ds.$$

By taking $y = \frac{t-s}{t-\tau}$, $dy = -\frac{ds}{t-\tau}$, $1 - y = \frac{s-\tau}{t-\tau}$ and limits $s : \tau \to t$, $y : 1 \to 0$

$$I_2 = \int_0^t f'(\tau) E_\alpha(A(t-\tau)^\alpha) d\tau.$$

Let

$$u = E_\alpha(A(t-\tau)^\alpha), \quad dv = f'(\tau) d\tau$$
$$du = -A(t-\tau)^{\alpha-1} E_{\alpha,\alpha}(A(t-\tau)^\alpha) d\tau, \quad v = f(\tau).$$

Then

$$I_2 = E_\alpha(A(t-\tau)^\alpha) f(\tau) \Big|_0^t + A \int_0^t (t-\tau)^{\alpha-1} E_{\alpha,\alpha}(A(t-\tau)^\alpha) f(\tau) d\tau$$

$$= f(t) - f(0) E_\alpha(At^\alpha) + A \int_0^t (t-\tau)^{\alpha-1} E_{\alpha,\alpha}(A(t-\tau)^\alpha) f(\tau) d\tau$$

and from (3.9),

$$L_2 = f(0) E_\alpha(At^\alpha) + f(t) - f(0) E_\alpha(At^\alpha)$$
$$+ A \int_0^t (t-s)^{\alpha-1} E_{\alpha,\alpha}(A(t-s)^\alpha) f(s) ds,$$

$$= f(t) + A \int_0^t (t-s)^{\alpha-1} E_{\alpha,\alpha}(A(t-s)^\alpha) f(s) ds.$$

From (3.8),

$$^C D^\alpha x(t) = L_1 + L_2$$
$$= A E_\alpha(At^\alpha) x_0 + f(t)$$
$$+ A \int_0^t (t-s)^{\alpha-1} E_{\alpha,\alpha}(A(t-s)^\alpha) f(s) ds$$
$$= A[E_\alpha(At^\alpha) x_0 + \int_0^t (t-s)^{\alpha-1} E_{\alpha,\alpha}(A(t-s)^\alpha) f(s) ds] + f(t)$$
$$= Ax(t) + f(t)$$

and $x(0) = x_0$.

Remark: Consider the linear fractional differential equation of the form

$$\left. \begin{array}{c} ^C D^\alpha x(t) + A^2 x(t) = f(t), \\ x(0) = x_0, \; x'(0) = y_0, \end{array} \right\} \tag{3.10}$$

where $1 < \alpha \le 2$, $x \in \mathbb{R}^n$, A is an $n \times n$ matrix and f is a continuous function. Applying Laplace transform to both sides, we get

$$s^\alpha X(s) - s^{\alpha-1}x(0) - s^{\alpha-2}x'(0) + A^2X(s) = F(s).$$

Then

$$X(s) = \frac{s^{\alpha-1}}{s^\alpha I + A^2}x_0 + \frac{s^{\alpha-2}}{s^\alpha I + A^2}y_0 + \frac{F(s)}{s^\alpha I + A^2}.$$

Taking inverse Laplace transform on both sides, we get

$$\mathcal{L}^{-1}\{X(s)\}(t) = \mathcal{L}^{-1}\left\{s^{\alpha-1}\left(s^\alpha I + A^2\right)^{-1}\right\}(t)x_0 + \mathcal{L}^{-1}\left\{s^{\alpha-2}\left(s^\alpha I + A^2\right)^{-1}\right\}(t)y_0$$

$$+\mathcal{L}^{-1}\left\{F(s)\left(s^\alpha I + A^2\right)^{-1}\right\}(t).$$

Using the Laplace transform of Mittag–Leffler function, we get the solution of the system (3.10) as

$$x(t) = E_\alpha(-A^2t^\alpha)x_0 + tE_{\alpha,2}(-A^2t^\alpha)y_0 + f(t) * t^{\alpha-1}E_{\alpha,\alpha}(-A^2t^\alpha)$$

$$= \Phi_0(t)x_0 + \Phi_1(t)y_0 + \int_0^t \Phi(t-s)f(s)ds,$$

where

$$\Phi_0(t) = E_\alpha(-A^2t^\alpha),$$
$$\Phi_1(t) = tE_{\alpha,2}(-A^2t^\alpha),$$
$$\Phi(t) = t^{\alpha-1}E_{\alpha,\alpha}(-A^2t^\alpha).$$

Note that when $\alpha = 2$, the linear fractional differential equation (3.10) reduces to the second order differential equation

$$\frac{d^2x(t)}{dt^2} + A^2x(t) = f(t),$$

with the same initial conditions $x(0) = x_0$ and $x'(0) = y_0$ and the solution takes the form

$$x(t) = \cos(At)x_0 + A^{-1}\sin(At)y_0 + \int_0^t A^{-1}\sin(A(t-s))f(s)ds.$$

3.4 Nonlinear Equations

Consider the following nonlinear fractional differential equation:

$$^C D^\alpha x(t) = Ax(t) + f(t, x(t)), \qquad (3.11)$$
$$x(t) = x_0,$$

where $0 < \alpha < 1, x \in \mathbb{R}^n$, A is an $n \times n$ matrix and $f : J \times \mathbb{R}^n \to \mathbb{R}^n$ is continuous. The integral representation of solution of Eq. (3.11) is given by

$$x(t) = E_\alpha(At^\alpha)x_0 + \int_0^t (t - s)^{\alpha-1} E_{\alpha,\alpha}(A(t - s)^\alpha) f(s, x(s)) ds. \qquad (3.12)$$

The existence and uniqueness of the solution of (3.11) implies the existence and uniqueness of the solution of (3.12) and vice versa. See [13] for details. Let $C(J, \mathbb{R}^n)$ denote the Banach space of continuous functions $x(t)$ with values in $\mathbb{R}^n$ for $t \in J$ with the norm

$$||x|| = \sup\{|x(t)| : t \in J\}.$$

Assume the following conditions:

(H_1) There exist constants $M_0 > 0$, $M > 0$ such that $||E_\alpha(At^\alpha)|| \le M_0$ and $||E_{\alpha,\alpha}(At^\alpha)|| \le M$.

(H_2) $f : J \times \mathbb{R}^n \to \mathbb{R}^n$ is continuous and there exist constants $L, N > 0$ such that

$$||f(t, x_1) - f(t, x_2)|| \le L||x_1 - x_2||, \quad \text{for all } x_1, x_2 \in \mathbb{R}^n.$$

and $N = \max ||f(t, 0)||$.

Theorem 3.4.1 *If the hypotheses (H_1) and (H_2) are satisfied and $M\frac{T^\alpha}{\alpha}L < 1$, then Eq. (3.11) has a unique solution on J.*

Proof Define the mapping $\Phi : C(J; \mathbb{R}^n) \to C(J; \mathbb{R}^n)$ by

$$\Phi x(t) = E_\alpha(At^\alpha)x_0 + \int_0^t (t - s)^{\alpha-1} E_{\alpha,\alpha}(A(t - s)^\alpha) f(s, x(s)) ds$$

and we need to show that Φ has a fixed point. This fixed point is the solution of Eq. (3.11). Choose

$$r \ge \frac{M_0||x_0|| + N}{1 - (M\frac{T^\alpha}{\alpha}L)}.$$

Let $B_r = \{x \in C(J; \mathbb{R}^n); ||x|| \le r\}$. Then we have to show that $\Phi B_r \subset B_r$. From the assumptions, we have

$$||\Phi x(t)|| \leq ||E_\alpha(At^\alpha)|| \, ||x_0|| + \int_0^t (t-s)^{\alpha-1}||E_{\alpha,\alpha}(A(t-s)^\alpha)|| \, ||f(s,x(s))|| ds,$$

$$= ||E_\alpha(At^\alpha)|| \, ||x_0 + \int_0^t (t-s)^{\alpha-1}||E_{\alpha,\alpha}(A(t-s)^{\alpha-1})||$$

$$[||f(s,x(s)) - f(s,0)|| + ||f(s,0)||] ds,$$

$$= M_0||x_0|| + \frac{T^\alpha}{\alpha} ML||x|| + N,$$

$$\leq r.$$

Thus, Φ maps B_r into itself. Now, for $x_1, x_2 \in C(J, \mathbb{R}^n)$, we have

$$||\Phi x_1(t) - \Phi x_2(t)|| \leq \int_0^t (t-s)^{\alpha-1}||E_{\alpha,\alpha}(A(t-s)^\alpha)|| L||x_1 - x_2|| ds,$$

$$\leq M \frac{T^\alpha}{\alpha} L||x_1 - x_2||,$$

$$\leq \frac{1}{2}||x_1 - x_2||.$$

Hence, Φ is a contraction mapping and therefore there exists a unique fixed point $x \in B_r$ such that $\Phi x(t) = x(t)$. Any fixed point of Φ is the solution of Eq. (3.11). $\square$

Theorem 3.4.2 *If the hypotheses (H_1) and (H_2) are satisfied, then Eq. (3.11) has a unique solution on J.*

Proof The proof is based on the application of fixed point method. Define a mapping $\Gamma : C(J, \mathbb{R}^n) \to C(J, \mathbb{R}^n)$ by

$$\Gamma x(t) = E_\alpha(At^\alpha)x_0 + \int_0^t (t-s)^{\alpha-1} E_{\alpha,\alpha}(A(t-s)^\alpha) f(s,x(s)) ds.$$

Let $x_1, x_2 \in C(J; \mathbb{R}^n)$. Then from Eq. (3.13), we have, for each $t \in J$

$$||\Gamma x_1(t) - \Gamma x_2(t)|| \leq M \frac{T^\alpha}{\alpha} L||x_1 - x_2||.$$

Then by induction, we have

$$||\Gamma^n x_1(t) - \Gamma^n x_2(t)|| \leq \frac{(M \frac{T^\alpha}{\alpha} L)^n}{n!} ||x_1 - x_2||.$$

Since $\frac{(M \frac{T^\alpha}{\alpha} L)^n}{n!} < 1$ for large n, by the generalization of the Banach contraction principle (Theorem 1.5.2), Γ has a unique fixed point $x \in C(J; \mathbb{R}^n)$. This fixed point is the solution of Eq. (3.11). $\square$

Now, we prove an interesting theorem which is very useful for proving the existence of solutions of nonlinear fractional differential equations in general Banach spaces.

Theorem 3.4.3 *Let* $X = \{x \in C([0, T]; \mathbb{R}) : \; {}^C D^\alpha x \in C([0, T]; \mathbb{R})\}$ *be the space equipped with the norm*

$$\|x\|_X = \max_{t \in [0,T]} |x(t)| + \max_{t \in [0,T]} |{}^C D^\alpha x|.$$

Then X is a Banach space.

Proof If $\alpha = 1$, then $X = C^1([0, T]; \mathbb{R})$ and in this case, X is a Banach space. Let us assume that $\alpha \in (0, 1)$. Let $\{x_n\} \subset X$ be a Cauchy sequence. Then $\{x_n\}$ and $\{{}^C D^\alpha x_n\}$ are Cauchy sequences in the space $C([0, T]; \mathbb{R})$, and so there exist $x, y \in C([0, T]; \mathbb{R})$ such that

$$\lim_{n \to \infty} x_n = x \Rightarrow \lim_{n \to \infty} \|x_n - x\| = 0$$

and

$$\lim_{n \to \infty} {}^C D^\alpha x_n = y \Rightarrow \lim_{n \to \infty} \|{}^C D^\alpha x_n - y\| = 0.$$

We want to prove that $y = {}^C D^\alpha x$; for that,

$$
\begin{aligned}
\int_0^t {}^C D^\alpha x_n(s) ds &= \int_0^t \frac{1}{\Gamma(1 - \alpha)} \int_0^s (s - \tau)^{-\alpha} x_n'(\tau) d\tau ds \\
&= \int_0^t \frac{ds}{\Gamma(1 - \alpha)} \int_0^s (s - \tau)^{-\alpha} x_n'(\tau) d\tau \\
&= \int_0^t \frac{x_n'(\tau) d\tau}{\Gamma(1 - \alpha)} \int_\tau^t (s - \tau)^{-\alpha} ds \\
&= \int_0^t \frac{x_n'(\tau) d\tau}{\Gamma(1 - \alpha)} \cdot \frac{(s - \tau)^{1-\alpha}}{1 - \alpha} \Big|_\tau^t \\
&= \int_0^t \frac{x_n'(\tau) d\tau}{\Gamma(1 - \alpha)} \frac{(t - \tau)^{1-\alpha}}{1 - \alpha} \\
&= \frac{1}{\Gamma(2 - \alpha)} \int_0^t (t - \tau)^{1-\alpha} x_n'(\tau) d\tau.
\end{aligned}
$$

Let

$$u = (t - \tau)^{1-\alpha}, \quad dv = x_n'(\tau) d\tau,$$
$$du = -(1 - \alpha)(t - \tau)^{-\alpha} d\tau, \quad v = x_n(\tau).$$

Then

$$\int_0^t {}^C D^\alpha x_n(s) ds = \frac{1}{\Gamma(2-\alpha)} \left[(t-\tau)^{1-\alpha} x_n(\tau) \Big|_0^t + \int_0^t (1-\alpha)(t-\tau)^{-\alpha} x_n(\tau) d\tau \right]$$

$$= \frac{-1}{\Gamma(2-\alpha)} t^{1-\alpha} x_n(0) + \frac{1}{\Gamma(1-\alpha)} \int_0^t (t-\tau)^{-\alpha} x_n(\tau) d\tau. \quad (3.13)$$

Consider

$$\frac{-1}{\Gamma(1-\alpha)} \int_0^t (t-\tau)^{-\alpha} x_n(0) d\tau = \frac{-x_n(0)}{\Gamma(1-\alpha)} \int_0^t (t-\tau)^{-\alpha} d\tau$$

$$= \frac{-x_n(0)}{\Gamma(1-\alpha)} \cdot \frac{-(t-\tau)^{1-\alpha}}{1-\alpha} \Big|_0^t$$

$$= \frac{-x_n(0)}{\Gamma(1-\alpha)} \left[-0 + \frac{t^{1-\alpha}}{1-\alpha} \right]$$

$$= \frac{-x_n(0)}{\Gamma(2-\alpha)} t^{1-\alpha}$$

and (3.13) implies

$$\int_0^t {}^C D^\alpha x_n(s) ds = \frac{1}{\Gamma(1-\alpha)} \int_0^t (t-\tau)^{-\alpha} (x_n(\tau) - x_n(0)) d\tau,$$

that is,

$$\int_0^t {}^C D^\alpha x_n(s) ds = \frac{1}{\Gamma(1-\alpha)} \int_0^t (t-s)^{-\alpha} (x_n(s) - x_n(0)) ds.$$

Letting $n \to \infty$ gives

$$\lim_{n\to\infty} \int_0^t {}^C D^\alpha x_n(s) ds = \lim_{n\to\infty} \frac{1}{\Gamma(1-\alpha)} \int_0^t (t-s)^{-\alpha} (x_n(s) - x_n(0)) ds.$$

By using the Lebesgue-dominated convergence theorem,

$$\int_0^t y(s) ds = \frac{1}{\Gamma(1-\alpha)} \int_0^t (t-s)^{-\alpha} (x(s) - x(0)) ds.$$

Taking differentiation with respect to t on both sides, we have

$$\frac{d}{dt} \int_0^t y(s) ds = \frac{1}{\Gamma(1-\alpha)} \frac{d}{dt} \int_0^t (t-s)^{-\alpha} (x(s) - x(0)) ds$$

$$y(t) = \frac{1}{\Gamma(1-\alpha)} \frac{d}{dt} \int_0^t (t-s)^{-\alpha} (x(s) - x(0)) ds$$

Let

$$u = x(s) - x(0), \quad dv = (t - s)^{-\alpha}ds$$
$$du = x'(s)ds, \quad v = -\frac{(t-s)^{1-\alpha}}{1-\alpha}.$$

Then

$$y(t) = \frac{1}{\Gamma(1-\alpha)} \frac{d}{dt} \left[-\frac{(t-s)^{1-\alpha}}{1-\alpha}(x(s)-x(0)) \Big|_0^t \right.$$
$$\left. + \int_0^t \frac{(t-s)^{1-\alpha}}{1-\alpha}x'(s)ds \right]$$

$$= \frac{1}{\Gamma(1-\alpha)} \frac{d}{dt} \left[-0.(x(t)-x(0)) + \frac{t^{1-\alpha}}{1-\alpha}.0 \right.$$
$$\left. + \int_0^t \frac{(t-s)^{1-\alpha}}{1-\alpha}x'(s)ds \right]$$

$$= \frac{1}{\Gamma(1-\alpha)} \frac{d}{dt} \int_0^t \frac{(t-s)^{1-\alpha}}{1-\alpha}x'(s)ds$$

$$= \frac{1}{\Gamma(1-\alpha)} \left[\int_0^t \frac{d}{dt}\frac{(t-s)^{1-\alpha}}{1-\alpha}x'(s)ds + 1.0 - 0 \right]$$

$$= \frac{1}{\Gamma(1-\alpha)} \int_0^t \frac{(1-\alpha)(t-s)^{-\alpha}}{(1-\alpha)}x'(s)ds$$

$$= \frac{1}{\Gamma(1-\alpha)} \int_0^t (t-s)^{-\alpha}x'(s)ds$$

$$= {}^C D^\alpha x(t).$$

Finally, we have

$$\lim_{n\to\infty} {}^C D^\alpha x_n(t) = y(t) = {}^C D^\alpha x(t).$$

Thus, the space X is a Banach space. $\qquad\qquad\square$

Consider the general fractional differential equation

$$^C D^\alpha x(t) = f(t, x(t)), \qquad (3.14)$$
$$x(0) = x_0,$$

where $0 < \alpha < 1$ and the nonlinear function $f : G \to \mathbb{R}$ is continuous. Here, $G = \{(t, x) : t \in [0, h^*], |x - x_0| < K, \text{ for some } K > 0, h^* > 0\}$ and

$$M = \sup_{(t,x)\in G} |f(t, x)|$$

and

$$h = \min\left\{h^*, \left(\frac{K\Gamma(\alpha+1)}{M}\right)^{\frac{1}{\alpha}}\right\}.$$

Now we prove that the initial value problem has a solution $x(t) \in C([0, h]; \mathbb{R})$. For that, we state the following result.

Lemma 3.4.4 *The function $x(t) \in C([0, h]; \mathbb{R})$ is a solution of the initial value problem (3.14) if and only if it is a solution of the integral equation*

$$x(t) = x_0 + \frac{1}{\Gamma(\alpha)} \int_0^t (t-s)^{\alpha-1} f(s, x(s)) ds. \tag{3.15}$$

For the sake of simplicity of the presentation, we only treat the scalar case explicitly here. However, all the results can be extended to vector valued functions without any difficulty.

Theorem 3.4.5 *The initial value problem (3.14) has a solution $x(t) \in C([0, h]; \mathbb{R})$*

Proof If $M = 0$, then $f(t, x) = 0$ for all $(t, x) \in G$. In this case, $x(t) = x_0$ is the solution of the initial value problem (3.14). When $M \neq 0$, the initial value problem (3.14) is equivalent to the integral equation (3.15). Set $U = \{x \in C([0, h] : \mathbb{R}) : \|x - x_0\| \leq K\}$. Obviously, U is a closed convex subset of the Banach space $C([0, h] : \mathbb{R})$ and so U is also a Banach space.

Define the operator $P : U \to U$ by

$$Px(t) = x_0 + \frac{1}{\Gamma(\alpha)} \int_0^t (t-s)^{\alpha-1} f(s, x(s)) ds. \tag{3.16}$$

First, we show that $Px \in U$ whenever $x \in U$. Now for $0 \leq t_1 \leq t_2 \leq h$,

$$|Px(t_1) - Px(t_2)|$$
$$= \frac{1}{\Gamma(\alpha)} \left| \int_0^{t_1} (t_1-s)^{\alpha-1} f(s, x(s)) ds - \int_0^{t_2} (t_2-s)^{\alpha-1} f(s, x(s)) ds \right|$$

$$= \frac{1}{\Gamma(\alpha)} \left| \int_0^{t_1} [(t_1-s)^{\alpha-1} - (t_2-s)^{\alpha-1}] f(s, x(s)) ds \right.$$
$$\left. + \int_{t_1}^{t_2} (t_2-s)^{\alpha-1} f(s, x(s)) ds \right|$$

$$\leq \frac{M}{\Gamma(\alpha)} \left(\int_0^{t_1} |(t_1-s)^{\alpha-1} - (t_2-s)^{\alpha-1}| ds + \int_{t_1}^{t_2} (t_2-s)^{\alpha-1} ds \right)$$

$$= \frac{M}{\Gamma(\alpha)} \int_0^{t_1} [(t_1 - s)^{\alpha-1} - (t_2 - s)^{\alpha-1}] ds + \int_{t_1}^{t_2} (t_2 - s)^{\alpha-1} ds,$$

$$= \frac{M}{\Gamma(\alpha)} \left[\frac{1}{\alpha} [t_1^\alpha - t_2^\alpha + (t_2 - t_1)^\alpha] + \frac{(t_2 - t_1)^\alpha}{\alpha} \right]$$

$$\leq \frac{2M}{\Gamma(\alpha + 1)} (t_2 - t_1)^\alpha, \text{ since } 0 < \alpha < 1,$$

which tends to zero as $t_2 \to t_1$ and so P is continuous. Moreover,

$$|Px(t) - x_0| = \frac{1}{\Gamma(\alpha)} \left| \int_0^t (t - s)^{\alpha-1} f(s, x(s)) ds \right|$$

$$\leq \frac{1}{\Gamma(\alpha + 1)} M t^\alpha$$

$$\leq \frac{M h^\alpha}{\Gamma(\alpha + 1)}$$

$$\leq \frac{M}{\Gamma(\alpha + 1)} \frac{K \Gamma(\alpha + 1)}{M} = K.$$

Thus, $Px \in U$ and so P maps U into U itself.

Next, we show that $P(U) = \{Px : x \in U\}$ is a relatively compact set. This can be done by means of the Arzela–Ascoli theorem. For $z \in P(U)$, $t \in [0, h]$,

$$|z(t)| = |Px(t)| \leq \|x_0\| + \frac{1}{\Gamma(\alpha)} \int_0^t (t - s)^{\alpha-1} |f(s, x(s))| ds$$

$$\leq \|x_0\| + \frac{1}{\Gamma(\alpha + 1)} M h^\alpha \leq \|x_0\| + K'$$

which is the required boundedness property. For equicontinuity, we have, for $0 \leq t_1 \leq t_2 \leq h$,

$$\|Px(t_1) - Px(t_2)\| \leq \frac{2M}{\Gamma(\alpha + 1)} (t_2 - t_1)^\alpha.$$

Thus, if $|t_1 - t_2| < \delta$, then

$$\|Px(t_1) - Px(t_2)\| \leq \frac{2M}{\Gamma(\alpha + 1)} \delta^\alpha.$$

Noting that the expression on the right-hand side is independent of x, t_1 and t_2, we see that the set $P(U)$ is equicontinuous. Hence, by the Arzela–Ascoli theorem, $P(U)$ is relatively compact and so, by Schauder's fixed point theorem, P has a fixed point. This fixed point is a solution of the initial value problem (3.14). □

3.5 Nonlinear Damped Equations

Consider the nonlinear fractional differential equation of the form

$$\left.\begin{array}{l} {}^{C}D^{\alpha}x(t) + A^2 x(t) = f(t, x(t), {}^{C}D^{\beta}x(t)), \ t \in J = [0, T], \\[6pt] x(0) = x_0, \ x'(0) = y_0, \end{array}\right\} \qquad (3.17)$$

with $1 < \alpha \leq 2$, $0 < \beta \leq 1$, A is an $n \times n$ matrix and the nonlinear function $f : J \times \mathbb{R}^n \times \mathbb{R}^n \to \mathbb{R}^n$ is continuous. The solution of (3.17) is given by

$$x(t) = \Phi_0(t) x_0 + \Phi_1(t) y_0 + \int_0^t \Phi(t - s) f(s, x(s), {}^{C}D^{\beta}x(s)) ds$$

where Φ, Φ_0, Φ_1 are already defined. For brevity, let us take

$$
\begin{aligned}
n_1 &= \sup\{\|\Phi_0(t)\|, t \in J\}; & n_2 &= \sup\{\|\Phi_1(t)\|, t \in J\}; \\
n_3 &= \sup\{\|\Phi(t - s)\|, t, s \in J\}; & n_4 &= \sup\{\|A^2\Phi(t)\|, t \in J\}; \\
n_5 &= \sup\{\|\Phi_2(t - s)\|, t, s \in J\}; & \Phi_2(t) &= t^{\alpha-1}E_{\alpha,\alpha-1}(-A^2 t^\alpha); \\
n_6 &= n_4\|x_0\| + n_1\|y_0\|; & c &= n_1\|x_0\| + n_2\|y_0\|.
\end{aligned}
$$

Now, we make the following assumptions to obtain the existence results for the equation:

(H1) For each $t \in J$, the function $f(t, \cdot, \cdot) : \mathbb{R}^n \times \mathbb{R}^n \to \mathbb{R}^n$ is continuous and the function $f(\cdot, x, y) : J \to \mathbb{R}^n$ is strongly measurable for each $x, y \in \mathbb{R}^n$.

(H2) For every positive constant k, there exists $h_k \in L^1(J)$ such that

$$\sup_{\|x\|, \|y\| \leq k} \|f(t, x, y)\| \leq h_k(t), \quad \text{for every } t \in J.$$

(H3) There exists a continuous function $m_1 : J \to [0, \infty)$ such that

$$\|f(t, x, y)\| \leq m_1(t)\Omega(\|x\| + \|y\|), \ t \in J, \ x, y \in \mathbb{R}^n,$$

where $\Omega : (0, \infty) \to (0, \infty)$ is a continuous nondecreasing function.

(H4) There exists a constant $M > 0$ and a continuous function $m_2 : J \to [0, \infty)$ such that

$$\frac{n_6 t^{-\beta}}{\Gamma(1 - \beta)} + \frac{n_5}{\Gamma(1 - \beta)} \int_0^t (t - s)^{-\beta} m_1(s)\Omega(w(s)) ds \leq M m_2(t)\Omega(w(t)),$$

and

$$\int_0^T m(s) ds < \int_c^\infty \frac{ds}{\Omega(s)}.$$

where

$$m(t) = \max\{n_3 m_1(t), M m_2(t)\}.$$

Theorem 3.5.1 *Assume that the hypotheses (H1)–(H4) hold. Then there exists a solution to the nonlinear equation (3.17) on J.*

Proof Consider the Banach space $X = \left\{x : x \in C(J, \mathbb{R}^n) \text{ and } {}^C D^\beta x \in C(J, \mathbb{R}^n)\right\}$ with norm $\|x\|^* = \max\{\|x\|, \|{}^C D^\beta x\|\}$. We now show that the nonlinear operator $F : X \to X$ defined by

$$(Fx)(t) = \Phi_0(t) x_0 + \Phi_1(t) y_0 + \int_0^t \Phi(t - s) f(s, x(s), {}^C D^\beta x(s)) ds$$

has a fixed point. This fixed point is then a solution to (3.17). Now

$$(Fx)(t) = \Phi_0(t) x_0 + \Phi_1(t) y_0 + \int_0^t \Phi(t - s) f(s, x(s), {}^C D^\beta x(s)) ds.$$

The first step is to obtain a priori bound of the set

$$\zeta(F) = \{x \in X : x = \lambda Fx \text{ for some } \lambda \in (0, 1)\}.$$

Let $x \in \zeta(F)$. Then $x = \lambda Fx$ for some $0 < \lambda < 1$. Thus, for each $t \in J$, we have

$$x(t) = \lambda \Phi_0(t) x_0 + \lambda \Phi_1(t) y_0 + \lambda \int_0^t \Phi(t - s) f(s, x(s), {}^C D^\beta x(s)) ds.$$

Then

$$\|x(t)\| \leq n_1 \|x_0\| + n_2 \|y_0\| + n_3 \int_0^t m_1(s) \Omega(\|x(s)\| + \|{}^C D^\beta x(s)\|) ds$$

$$\equiv c + n_3 \int_0^t m_1(s) \Omega(\|x(s)\| + \|{}^C D^\beta x(s)\|) ds.$$

Denoting the right-hand side of the above inequality by $r_1(t)$, we have $r_1(0) = c$,

$$\|x(t)\| \leq r_1(t)$$

and

$$r_1'(t) = n_3 m_1(t) \Omega(\|x(t)\| + \|{}^C D^\beta x(t)\|).$$

Also,

$$x'(t) = -\lambda A^2 \Phi(t) x_0 + \lambda \Phi_0(t) y_0 + \lambda \int_0^t \Phi_2(t-s) f(s, x(s), {}^C D^\beta x(s)) ds.$$

and

$$
\begin{aligned}
\|x'(t)\| &\leq n_4 \|x_0\| + n_1 \|y_0\| + n_5 \int_0^t m_1(s)\Omega(\|x(s)\| + \|{}^C D^\beta x(s)\|) ds \\
&\equiv n_6 + n_5 \int_0^t m_1(s)\Omega(\|x(s)\| + \|{}^C D^\beta x(s)\|) ds.
\end{aligned}
$$

Hence, it follows that

$$
\begin{aligned}
\|{}^C D^\beta x(t)\| &\leq \frac{1}{\Gamma(1-\beta)} \int_0^t (t-s)^{-\beta} \|x'(s)\| ds \\
&\leq \frac{n_6}{\Gamma(1-\beta)} \int_0^t (t-s)^{-\beta} ds \\
&\quad + \frac{n_5}{\Gamma(1-\beta)} \int_0^t (t-s)^{-\beta} \left(\int_0^s m_1(\tau)\Omega(\|x(\tau)\| + \|{}^C D^\beta x(\tau)\|) d\tau \right) ds \\
&\leq \frac{n_6}{\Gamma(1-\beta)} \int_0^t (t-s)^{-\beta} ds \\
&\quad + \frac{n_5}{\Gamma(1-\beta)} \int_0^t \int_\tau^t (t-s)^{-\beta} ds\, m_1(\tau)\Omega(\|x(\tau)\| + \|{}^C D^\beta x(\tau)\|) d\tau \\
&\leq \frac{n_6 t^{1-\beta}}{\Gamma(2-\beta)} + \frac{n_5}{\Gamma(2-\beta)} \int_0^t (t-\tau)^{1-\beta} m_1(\tau)\Omega(\|x(\tau)\| + \|{}^C D^\beta x(\tau)\|) d\tau.
\end{aligned}
$$

Denoting the right-hand side of the above inequality by $r_2(t)$, we have $r_2(0) = 0$ and

$$\|{}^C D^\beta x(t)\| \leq r_2(t)$$

and

$$r_2'(t) = \frac{n_6 t^{-\beta}}{\Gamma(1-\beta)} + \frac{n_5}{\Gamma(1-\beta)} \int_0^t (t-\tau)^{-\beta} m_1(\tau)\Omega(\|x(\tau)\| + \|{}^C D^\beta x(\tau)\|) d\tau.$$

Let $w(t) = r_1(t) + r_2(t)$, $t \in J$. Then $w(0) = r_1(0) + r_2(0) = c$ and

$$w'(t) = r_1'(t) + r_2'(t) \leq m(t)\Omega(w(t))$$

which implies that for each $t \in J$,

$$\int_{w(0)}^{w(t)} \frac{ds}{\Omega(s)} \leq \int_c^T m(s) ds < \int_c^\infty \frac{ds}{\Omega(s)}.$$

From the above inequality, we see that there exists a constant K such that

$$w(t) = r_1(t) + r_2(t) \leq K, \ t \in J.$$

Then $\|x(t)\| \leq r_1(t)$ and $\|{}^C D^\beta x(t)\| \leq r_2(t), \ t \in J$, and hence

$$\|x\|^* = \max\{\|x\|, \|{}^C D^\beta x\|\} \leq K$$

and the set $\zeta(F)$ is bounded. Next, we prove that the operator $F : X \to X$ is completely continuous.

Let $B_q = \{x \in X : \|x\|^* \leq q\}$. We first show that F maps bounded sets into equicontinuous family in B_q. Let $x \in B_q$ and $t_1, t_2 \in J$. Then if $0 < t_1 < t_2 \leq T$,

$$
\begin{aligned}
\|(Fx)(t_2) - (Fx)(t_1)\| \ \leq \ & \|\Phi_0(t_2) - \Phi_0(t_1)\|\|x_0\| + \|\Phi_1(t_2) - \Phi_1(t_1)\|\|y_0\| \\
& + \left\| \int_0^{t_1} [\Phi(t_2 - s) - \Phi(t_1 - s)] f(s, x(s), {}^C D^\beta x(s)) ds \right\| \\
& + \left\| \int_{t_1}^{t_2} \Phi(t_2 - s) f(s, x(s), {}^C D^\beta x(s)) ds \right\| \\
\leq \ & \|\Phi_0(t_2) - \Phi_0(t_1)\|\|x_0\| + \|\Phi_1(t_2) - \Phi_1(t_1)\|\|y_0\| \\
& + \int_0^{t_1} \|\Phi(t_2 - s) - \Phi(t_1 - s)\| h_q(s) ds \\
& + \int_{t_1}^{t_2} \|\Phi(t_2 - s)\| h_q(s) ds;
\end{aligned}
\tag{3.18}
$$

and

$$
\begin{aligned}
\|(Fx)'(t)\| \ \leq \ & \|A^2 \Phi(t)\|\|x_0\| + \|\Phi_0(t)\|\|y_0\| + \int_0^t \|\Phi_2(t - s)\| h_q(s) ds. \\
\leq \ & n_4\|x_0\| + n_1\|y_0\| + n_5 \int_0^t h_q(s) ds \\
\leq \ & n_6 + n_5 \int_0^t h_q(s) ds.
\end{aligned}
$$

Hence, it follows that

$$
\begin{aligned}
\|{}^C D^\beta (Fx)(t_2) & - {}^C D^\beta (Fx)(t_1)\| \\
= \ & \left\| \frac{1}{\Gamma(1 - \beta)} \int_0^{t_2} (t_2 - s)^{-\beta} (Fx)'(s) ds - \frac{1}{\Gamma(1 - \beta)} \int_0^{t_1} (t_1 - s)^{-\beta} (Fx)'(s) ds \right\| \\
\leq \ & \frac{1}{\Gamma(1 - \beta)} \left\| \int_{t_1}^{t_2} (t_2 - s)^{-\beta} (Fx)'(s) ds \right\| \\
& + \frac{1}{\Gamma(1 - \beta)} \left\| \int_0^{t_1} \left((t_2 - s)^{-\beta} - (t_1 - s)^{-\beta} \right) (Fx)'(s) ds \right\|
\end{aligned}
$$

$$\leq \frac{1}{\Gamma(1-\beta)} \int_{t_1}^{t_2} (t_2 - s)^{-\beta} \|(Fx)'(s)\| ds$$

$$+ \frac{1}{\Gamma(1-\beta)} \int_0^{t_1} \left((t_2 - s)^{-\beta} - (t_1 - s)^{-\beta} \right) \|(Fx)'(s)\| ds$$

$$\leq \frac{n_6}{\Gamma(1-\beta)} \int_{t_1}^{t_2} (t_2 - s)^{-\beta} ds + \frac{n_5}{\Gamma(1-\beta)} \int_{t_1}^{t_2} (t_2 - s)^{-\beta} \left(\int_0^s h_q(\tau) d\tau \right) ds$$

$$+ \frac{n_6}{\Gamma(1-\beta)} \int_0^{t_1} \left((t_2 - s)^{-\beta} - (t_1 - s)^{-\beta} \right) ds$$

$$+ \frac{n_5}{\Gamma(1-\beta)} \int_0^{t_1} \left((t_2 - s)^{-\beta} - (t_1 - s)^{-\beta} \right) \left(\int_0^s h_q(\tau) d\tau \right) ds$$

$$\leq \frac{n_6}{\Gamma(2-\beta)} (t_2^{1-\beta} - t_1^{1-\beta}) + \frac{1}{\Gamma(1-\beta)} \int_{t_1}^{t_2} (t_2 - s)^{-\beta} \left(\int_0^s h_q(\tau) d\tau \right) ds$$

$$+ \frac{1}{\Gamma(2-\beta)} \int_0^{t_1} \left((t_2 - \tau)^{1-\beta} - (t_2 - t_1)^{1-\beta} - (t_1 - \tau)^{1-\beta} \right) h_q(\tau) d\tau. \quad (3.19)$$

The right-hand sides of (3.18) and (3.19) tend to zero as $t_2 \to t_1$. Thus, F maps B_q into an equicontinuous family of functions. It is easy to see that the family FB_q is uniformly bounded. Next, we show that F is a compact operator. It suffices to show that the closure of FB_q is compact.

Let $0 \leq t \leq T$ be fixed and ϵ be a real number satisfying $0 < \epsilon < t$. For $x \in B_q$, we define

$$(F_\epsilon x)(t) = \Phi_0(t)x_0 + \Phi_1(t)y_0 + \int_0^{t-\epsilon} \Phi(t-s)f(s, x(s), {}^C D^\beta x(s)) ds.$$

Note that using the same methods as in the procedure above, we obtain the boundedness and equicontinuous property of F_ϵ which implies that the set $S_\epsilon(t) = \{(F_\epsilon x)(t) : x \in B_q\}$ is relatively compact in X for every $0 < \epsilon < t$.

Moreover, for every $x \in B_q$,

$$\|(Fx)(t) - (F_\epsilon x)(t)\| \leq \left\| \int_{t-\epsilon}^t \Phi(t-s)f(s, x(s), {}^C D^\beta x(s)) ds \right\|$$

$$\leq \int_{t-\epsilon}^t \|\Phi(t-s)\| h_q(s) ds.$$

Also

$$\|(Fx)'(t) - (F_\epsilon x)'(t)\| \leq \left\| \int_{t-\epsilon}^t \Phi_2(t-s)f(s, x(s), {}^C D^\beta x(s)) ds \right\|$$

$$\leq \int_{t-\epsilon}^t \|\Phi_2(t-s)\| h_q(s) ds.$$

Since $\|(Fx)(t) - (F_\epsilon x)(t)\| \to 0$ and $\|(Fx)'(t) - (F_\epsilon x)'(t)\| \to 0$ as $\epsilon \to 0$, this implies that

$$\|{}^{C}D^{\beta}(Fx)(t) \; - \; {}^{C}D^{\beta}(F_{\epsilon}x)(t)\|$$

$$\leq \; \frac{1}{\Gamma(1-\beta)} \int_{0}^{t} (t-s)^{-\beta} \|(Fx)'(t) - (F_{\epsilon}x)'(t)\| ds \; \to 0 \text{ as } \epsilon \to 0.$$

So relatively compact sets $S_{\epsilon}(t) = \{(F_{\epsilon}x)(t) : x \in B_{q}\}$ are arbitrarily close to the set $\{(Fx)(t) : x \in B_{q}\}$. Hence, $\{(Fx)(t) : x \in B_{q}\}$ is compact in X by the Arzela–Ascoli theorem.

Next, it remains to show that F is continuous. Let $\{x_{n}\}$ be a sequence in X such that $\|x_{n} - x\| \to 0$ as $n \to \infty$. Then there is an integer k such that $\|x_{n}\| \leq k$, $\|D^{\beta}x_{n}\| \leq k$ for all n and $t \in J$. So, $\|x(t)\| \leq k$, $\|{}^{C}D^{\beta}x(t)\| \leq k$ and x, $D^{\beta}x \in X$. By $(H1)$,

$$f(t, x_{n}(t), {}^{C}D^{\beta}x_{n}(t)) \to f(t, x(t), {}^{C}D^{\beta}x(t)),$$

for each $t \in J$. Since

$$\|f(t, x_{n}(t), {}^{C}D^{\beta}x_{n}(t)) - f(t, x(t), {}^{C}D^{\beta}x(t))\| \leq 2h_{k}(t),$$

we have, by the dominated convergence theorem,

$$\|(Fx_{n})(t) \; - \; (Fx)(t)\|$$

$$= \; \sup_{t \in J} \left\| \int_{0}^{t} \Phi(t-s) \left[f(s, x_{n}(s), {}^{C}D^{\beta}x_{n}(s)) - f(s, x(s), {}^{C}D^{\beta}x(s)) \right] ds \right.$$

$$\leq \; \int_{0}^{T} \left\| \Phi(t-s) \left[f(s, x_{n}(s), {}^{C}D^{\beta}x_{n}(s)) - f(s, x(s), {}^{C}D^{\beta}x(s)) \right] \right\| ds.$$

Also

$$\|(Fx_{n})'(t) \; - \; (Fx)'(t)\|$$

$$= \; \sup_{t \in J} \left\| \int_{0}^{t} \Phi_{2}(t-s) \left[f(s, x_{n}(s), {}^{C}D^{\beta}x_{n}(s)) - f(s, x(s), {}^{C}D^{\beta}x(s)) \right] ds \right.$$

$$\leq \; \int_{0}^{T} \left\| \Phi_{2}(t-s) \left[f(s, x_{n}(s), {}^{C}D^{\beta}x_{n}(s)) - f(s, x(s), {}^{C}D^{\beta}x(s)) \right] \right\| ds.$$

This implies that

$$\|{}^{C}D^{\beta}(Fx_{n})(t) \; - \; {}^{C}D^{\beta}(Fx)(t)\|$$

$$\leq \; \frac{1}{\Gamma(1-\beta)} \int_{0}^{t} (t-s)^{-\beta} \|(Fx_{n})'(t) - (Fx)'(t)\| ds \; \to 0 \text{ as } n \to \infty.$$

Thus, F is continuous. Finally, the set $\zeta(F) = \{x \in X : x = \lambda Fx, \ \lambda \in (0, 1)\}$ is bounded as shown in the first step. By Schaefer's theorem, the operator F has a fixed point in X. This fixed point is then the solution of (3.17).   □

3.6 Examples

Example 3.6.1 Consider the fractional differential equation in $\mathbb{R}$ and take $f = 0$, $x(0) = 1$, $A = 1$ in (3.3), we get

$$^C D^\alpha x(t) = x(t), t \in [0, T], \tag{3.20}$$
$$x(0) = 1,$$

where $0 < \alpha < 1$. The corresponding integral solution of (3.20) is

$$x(t) = 1 + \frac{1}{\Gamma(\alpha)} \int_0^t \frac{x(s)}{(t-s)^{1-\alpha}} ds. \tag{3.21}$$

The exact solution of (3.20) is given by

$$x(t) = E_\alpha(t^\alpha) = \sum_{k=0}^{\infty} \frac{(t^\alpha)^k}{\Gamma(\alpha k + 1)} \tag{3.22}$$

where E_α is the Mittag–Leffler function. On expanding, we have

$$x(t) = \frac{(t^\alpha)^0}{\Gamma(\alpha.0 + 1)} + \frac{(t^\alpha)^1}{\Gamma(\alpha + 1)} + \frac{(t^\alpha)^2}{\Gamma(2\alpha + 1)} + \cdots$$
$$= 1 + \frac{t^\alpha}{\Gamma(\alpha + 1)} + \frac{t^{2\alpha}}{\Gamma(2\alpha + 1)} + \cdots.$$

We want to show that both the solutions are same. For that using (3.22) in (3.21), we get

$$1 + \frac{t^\alpha}{\Gamma(\alpha + 1)} + \frac{t^{2\alpha}}{\Gamma(2\alpha + 1)} + \cdots$$

$$= 1 + \frac{1}{\Gamma(\alpha)} \int_0^t (t-s)^{\alpha-1} \left(1 + \frac{s^\alpha}{\Gamma(\alpha + 1)} + \frac{s^{2\alpha}}{\Gamma(2\alpha + 1)} + \cdots\right) ds$$
$$= 1 + \frac{1}{\Gamma(\alpha)} \left[\frac{-(t-s)^\alpha}{\alpha}\right]_0^t + \frac{1}{\Gamma(\alpha)\Gamma(\alpha + 1)} \int_0^t (t-s)^{\alpha-1} s^\alpha ds + \cdots$$
$$= 1 + \frac{t^\alpha}{\Gamma(\alpha + 1)} + \frac{1}{\Gamma(\alpha)\Gamma(\alpha + 1)} \int_0^t (t-s)^{\alpha-1} s^\alpha ds + \cdots.$$

Next, we prove that

$$\frac{t^{2\alpha}}{\Gamma(2\alpha+1)} = \frac{1}{\Gamma(\alpha)\Gamma(\alpha+1)} \int_0^t (t-s)^{\alpha-1} s^\alpha ds.$$

The value of the integral

$$\int_0^t (t-s)^{\alpha-1} s^\alpha ds = t^{\alpha-1} \int_0^t (1-\frac{s}{t})^{\alpha-1} s^\alpha ds,$$

$$= t^{\alpha-1} \int_0^1 (1-x)^{\alpha-1} (tx)^\alpha t dx, \text{ (taking } x = \frac{s}{t}),$$

$$= t^{\alpha-1} \int_0^1 (1-x)^{\alpha-1} t^{\alpha+1} x^\alpha dx,$$

$$= t^{2\alpha} \int_0^1 x^{\alpha+1-1} (1-x)^{\alpha-1} dx,$$

$$= t^{2\alpha} B(\alpha+1, \alpha),$$

$$= t^{2\alpha} \frac{\Gamma(\alpha+1)\Gamma(\alpha)}{\Gamma(2\alpha+1)}.$$

From the above, we get

$$\frac{1}{\Gamma(\alpha+1)\Gamma(\alpha)} \int_0^t (t-s)^{\alpha-1} s^\alpha ds = \frac{t^{2\alpha}}{\Gamma(2\alpha+1)}.$$

Similarly, we evaluate the other terms on the right-hand side of the series and it is equal to

$$1 + \frac{t^\alpha}{\Gamma(\alpha+1)} + \frac{t^{2\alpha}}{\Gamma(2\alpha+1)} + \cdots.$$

Hence, the solution representations (3.21) and (3.22) are the same.

Example 3.6.2 Consider the fractional differential equation with $f(t) = t, x(0) = 1, A = 1$, then

$$^C D^\alpha x(t) = x(t) + t, \ t \in [0, T] \tag{3.23}$$

$$x(0) = 1,$$

where $0 < \alpha < 1$. The corresponding integral solution of (3.23) is

$$x(t) = 1 + \frac{1}{\Gamma(\alpha)} \int_0^t \frac{x(s)}{(t-s)^{1-\alpha}} ds + \frac{1}{\Gamma(\alpha)} \int_0^t \frac{s}{(t-s)^{1-\alpha}} ds. \tag{3.24}$$

The exact solution of (3.23) is

$$x(t) = E_{\alpha,1}(t^\alpha) + \int_0^t (t - s)^{\alpha-1}[E_{\alpha,\alpha}(t - s)^\alpha]s\,ds. \tag{3.25}$$

On expanding

$$x(t) = 1 + \frac{t^\alpha}{\Gamma(\alpha + 1)} + \frac{t^{2\alpha}}{\Gamma(2\alpha + 1)} + \cdots$$
$$+ \int_0^t (t - s)^{\alpha-1}\left(\frac{1}{\Gamma(\alpha)} + \frac{(t - s)^\alpha}{\Gamma(2\alpha)} + \frac{(t - s)^{2\alpha}}{\Gamma(3\alpha)} + \cdots\right) s\,ds.$$

Using similar procedure as in the above example with judicious integral evaluation, we can show that both the solutions (3.24) and (3.25) for (3.23) are the same.

Example 3.6.3 For the Riemann–Liouville fractional differential equation

$$D^{\frac{4}{3}}x(t) = 0,$$

the solution is

$$x(t) = c_1 t^{\frac{1}{3}} + c_2 t^{-\frac{2}{3}}, \quad t > 0,$$

where c_1 and c_2 are arbitrary constants.

Example 3.6.4 Consider the equation

$$tD^{\frac{1}{2}}x(t) = x(t).$$

Taking Laplace transform, we get

$$D[s^{\frac{1}{2}}X(s) - D^{-\frac{1}{2}}x(0)] + X(s) = 0.$$

Differentiation gives

$$DX(s) + (\frac{1}{2}s^{-1} + s^{-\frac{1}{2}})X(s) = 0$$

which is a first-order differential equation in $X(s)$ and the solution is

$$X(s) = ks^{-\frac{1}{2}}e^{-2\sqrt{s}},$$

where k is a constant. By taking the inverse Laplace transform, we have

$$x(t) = \frac{k}{\sqrt{t}}e^{-\frac{1}{t}}, \quad t > 0.$$

Example 3.6.5 Basset Equation

Consider the Basset equation

$$\dot{x}(t) + a^C D^\alpha x(t) + x(t) = f(t), \quad t \in J = [0, T],$$

where $f(t)$ is a continuous function on J. Let $\alpha = \frac{1}{2}$ and $a \in \mathbb{R}$. Take $0 < a < 2$ and $a \neq 2$, the solution is

$$x(t) = \frac{1}{\lambda_1 - \lambda_2} \Big\{ [\lambda_1 E_{\frac{1}{2}}(\lambda_2 t^{\frac{1}{2}}) - \lambda_2 E_{\frac{1}{2}}(\lambda_1 t^{\frac{1}{2}})]x_0$$

$$+ \int_0^t (t-s)^{-\frac{1}{2}} [E_{\frac{1}{2},\frac{1}{2}}(\lambda_1 (t-s)^{\frac{1}{2}}) - E_{\frac{1}{2},\frac{1}{2}}(\lambda_2 (t-s)^{\frac{1}{2}})] f(s) ds \Big\},$$

where

$$\lambda_1 = \frac{-a + \sqrt{a^2 - 4}}{2},$$

$$\lambda_2 = \frac{-a - \sqrt{a^2 - 4}}{2}.$$

Now let us consider $a = 2$ and

$$\lambda_1 = \lambda_2 = -\frac{a}{2} = \lambda = -1$$

The polynomial

$$s + as^{\frac{1}{2}} + 1 = (s^{\frac{1}{2}} - \lambda)^2$$

$$= (s^{\frac{1}{2}} + 1)^2.$$

Now we consider

$$\frac{1}{s^{\frac{1}{2}}(s^{\frac{1}{2}} \pm \lambda)^2} = \mathcal{L}\Big\{ 2\sqrt{\frac{t}{\pi}} \mp 2\lambda t E_{\frac{1}{2}}(\mp \lambda t^\alpha) \Big\}(s), \quad \lambda \in \mathbb{C}.$$

Case (i): Now

$$\frac{1}{s^{\frac{1}{2}}(s^{\frac{1}{2}} \pm \lambda)^2} = \mathcal{L}\Big\{ 2\sqrt{\frac{t}{\pi}} - 2\lambda t E_{\frac{1}{2}}(-\lambda t^{\frac{1}{2}}) \Big\}(s)$$

$$\mathcal{L}\Big\{ 2\sqrt{\frac{t}{\pi}} \Big\}(s) = \int_0^\infty e^{-st} \frac{2\sqrt{t}}{\sqrt{\pi}} dt$$

$$= \frac{2}{\sqrt{\pi}} \int_0^\infty e^{-st} \sqrt{t} dt$$

$$\mathcal{L}\Big\{ 2\sqrt{\frac{t}{\pi}} \Big\}(s) = \frac{2}{\sqrt{\pi} s^{\frac{3}{2}}} \Gamma(\frac{3}{2})$$

$$\mathcal{L}\Big\{ 2\sqrt{\frac{t}{\pi}} \Big\}(s) = \frac{1}{s^{\frac{3}{2}}} \tag{3.26}$$

$$\mathcal{L}\left\{-2\lambda t E_{\frac{1}{2}}(-\lambda t^{\frac{1}{2}})\right\}(s) = \int_0^\infty e^{-st} - 2\lambda t E_{\frac{1}{2}}(-\lambda t^{\frac{1}{2}})dt$$

$$= -2\lambda \int_0^\infty e^{-st} t E_{\frac{1}{2}}(-\lambda t^{\frac{1}{2}})dt$$

$$= -2\lambda \int_0^\infty e^{-st} t \sum_{k=0}^\infty \frac{(-\lambda)^k t^{\frac{k}{2}}}{\Gamma\frac{k}{2}+1} dt$$

$$\mathcal{L}\left\{-2\lambda t E_{\frac{1}{2}}(-\lambda t^{\frac{1}{2}})\right\}(s) = \frac{1}{s^{\frac{1}{2}}(s^{\frac{1}{2}}+\lambda)^2} - \frac{1}{s^{\frac{3}{2}}}. \tag{3.27}$$

From (3.26) and (3.27), we have

$$\frac{1}{s^{\frac{1}{2}}(s^{\frac{1}{2}}\pm\lambda)^2} = \mathcal{L}\left\{2\sqrt{\frac{t}{\pi}} - 2\lambda t E_{\frac{1}{2}}(-\lambda t^{\frac{1}{2}})\right\}(s).$$

Case (ii): Now

$$\frac{1}{s^{\frac{1}{2}}(s^{\frac{1}{2}}-\lambda)^2} = \mathcal{L}\left\{2\sqrt{\frac{t}{\pi}} + 2\lambda t E_{\frac{1}{2}}(\lambda t^{\frac{1}{2}})\right\}(s),$$

we conclude that

$$\frac{1}{s^{\frac{1}{2}}(s^{\frac{1}{2}}\pm\lambda)^2} = \mathcal{L}\left\{2\sqrt{\frac{t}{\pi}} \mp 2\lambda t E_{\frac{1}{2}}(\mp\lambda t^{\frac{1}{2}})\right\}(s).$$

Next

$$\frac{1}{(s^{\frac{1}{2}}\pm\lambda)^2} = \mathcal{L}\left\{\mp 2\lambda \sqrt{\frac{t}{\pi}} + (1+2\lambda^2 t)E_{\frac{1}{2}}(\mp\lambda t^{\frac{1}{2}})\right\}(s).$$

Case (i): Now

$$\frac{1}{(s^{\frac{1}{2}}+\lambda)^2} = \mathcal{L}\left\{-2\lambda\sqrt{\frac{t}{\pi}} + (1+2\lambda^2 t)E_{\frac{1}{2}}(-\lambda t^{\frac{1}{2}})\right\}(s)$$

$$\mathcal{L}\left\{-2\lambda\sqrt{\frac{t}{\pi}}\right\}(s) = \frac{-\lambda}{s^{\frac{3}{2}}} \tag{3.28}$$

$$\mathcal{L}\left\{(1+2\lambda^2 t)E_{\frac{1}{2}}(-\lambda t^{\frac{1}{2}})\right\}(s) = \mathcal{L}\left\{E_{\frac{1}{2}}(-\lambda t^{\frac{1}{2}})\right\}(s)$$

$$+2\lambda^2 E_{\frac{1}{2}}(-\lambda t^{\frac{1}{2}})\right\}(s)$$

$$\mathcal{L}\left\{(1+2\lambda^2 t)E_{\frac{1}{2}}(-\lambda t^{\frac{1}{2}})\right\}(s) = \frac{1}{s^{\frac{1}{2}}(s^{\frac{1}{2}}+\lambda)^2} + \frac{\lambda}{s^{\frac{3}{2}}}. \tag{3.29}$$

From (3.28) and (3.29), we get

$$\frac{1}{(s^{\frac{1}{2}} + \lambda)^2} = \mathcal{L}\left\{ -2\sqrt{\frac{t}{\pi}} + (1 + 2\lambda^2 t) E_{\frac{1}{2}}(-\lambda t^{\frac{1}{2}}) \right\}(s)$$

Case (ii):

$$\frac{1}{(s^{\frac{1}{2}} - \lambda)^2} = \mathcal{L}\left\{ 2\sqrt{\frac{t}{\pi}} + (1 + 2\lambda^2 t) E_{\frac{1}{2}}(\lambda t^{\frac{1}{2}}) \right\}(s)$$

we conclude that

$$\frac{1}{(s^{\frac{1}{2}} \pm \lambda)^2} = \mathcal{L}\left\{ \mp 2\sqrt{\frac{t}{\pi}} + (1 + 2\lambda^2 t) E_{\frac{1}{2}}(\mp\lambda t^{\frac{1}{2}}) \right\}(s)$$

Consider the equation

$$\dot{x}(t) + 2^C D^{\frac{1}{2}} x(t) + x(t) = f(t),$$
$$x(0) = x_0. \tag{3.30}$$

Applying Laplace transform on both sides to get

$$sX(s) - x(0) + 2s^{\frac{1}{2}} X(s) - 2s^{-\frac{1}{2}} x(0) + X(s) = F(s).$$

On grouping, we get

$$X(s)(s + 2s^{\frac{1}{2}} + 1) = x(0) + 2s^{-\frac{1}{2}} x(0) + F(s)$$

$$X(s) = \frac{1 + 2s^{-\frac{1}{2}}}{s + 2s^{\frac{1}{2}} + 1} x_0 + \frac{F(s)}{s + 2s^{\frac{1}{2}} + 1}$$

$$\frac{1}{(s^{\frac{1}{2}} + 1)^2} = \mathcal{L}\left\{ -2\sqrt{\frac{t}{\pi}} + (1 + 2t) E_{\frac{1}{2}}(-t^{\frac{1}{2}}) \right\}(s)$$

$$\frac{1}{(s^{\frac{1}{2}} + 1)^2} = \mathcal{L}\left\{ -2\sqrt{\frac{t}{\pi}} - 2t E_{\frac{1}{2}}(-t^{\frac{1}{2}}) \right\}(s)$$

$$\frac{F(s)}{(s^{\frac{1}{2}} + 1)^2} = \mathcal{L}\left\{ f(t) * -2\sqrt{\frac{t}{\pi}} + (1 + 2t) E_{\frac{1}{2}}(-t^{\frac{1}{2}}) \right\}(s).$$

Taking inverse Laplace transform on both sides, we get

$$x(t) = \frac{2\sqrt{t}}{\sqrt{\pi}}x_0 + (1 - 2t)E_{\frac{1}{2}}(-\sqrt{t})x_0 - \frac{2}{\sqrt{\pi}}\int_0^t \sqrt{t - s}f(s)ds$$

$$+ \int_0^t (1 + 2(t - s))E_{\frac{1}{2}}(-\sqrt{(t - s)})f(s)ds.$$

Example 3.6.6 Consider the fractional differential equation

$$^CD^{3/2}x_1(t) - 2x_1(t) + 3x_2(t) = \frac{x_1}{x_1^2 + \sin t} ,$$

$$^CD^{3/2}x_2(t) - 4x_1(t) + 5x_2(t) = \frac{x_1}{x_1^2 + t} , \tag{3.31}$$

with initial conditions $\begin{bmatrix} x_1(0) \\ x_2(0) \end{bmatrix} = \begin{bmatrix} 1 \\ 1 \end{bmatrix}$ and $\begin{bmatrix} x_1'(0) \\ x_2'(0) \end{bmatrix} = \begin{bmatrix} 0 \\ 0 \end{bmatrix}$ for $t \in [0, 2]$.

It has the following form

$$^CD^{3/2}x(t) + A^2x(t) = f(t, x), \ t \in [0, 2],$$

$$x(0) = x_0, \ x'(0) = y_0, \tag{3.32}$$

where $A^2 = \begin{bmatrix} -2 & 3 \\ -4 & 5 \end{bmatrix}$, $f(t, x) = \begin{bmatrix} \dfrac{x_1}{x_1^2 + \sin t} \\ \dfrac{x_2}{x_2^2 + t} \end{bmatrix}$, $x(t) = \begin{bmatrix} x_1(t) \\ x_2(t) \end{bmatrix}$.

Using Mittag–Leffler matrix function for a given matrix A^2, we get

$$\Phi(2 - s) = \begin{bmatrix} L_1(s) & L_2(s) \\ L_3(s) & L_4(s) \end{bmatrix},$$

where

$$L_1(s) = (2 - s)^{1/2} \left[4E_{3/2,3/2}(-(2 - s)^{3/2}) - 3E_{3/2,3/2}(-2(2 - s)^{3/2}) \right],$$

$$L_2(s) = (2 - s)^{1/2} \left[3E_{3/2,3/2}(-2(2 - s)^{3/2}) - 3E_{3/2,3/2}(-(2 - s)^{3/2}) \right],$$

$$L_3(s) = (2 - s)^{1/2} \left[4E_{3/2,3/2}(-(2 - s)^{3/2}) - 4E_{3/2,3/2}(-2(2 - s)^{3/2}) \right],$$

$$L_4(s) = (2 - s)^{1/2} \left[4E_{3/2,3/2}(-2(2 - s)^{3/2}) - 3E_{3/2,3/2}(-(2 - s)^{3/2}) \right].$$

Further, the nonlinear function f is bounded, continuous, and satisfies conditions of Theorem 3.5.1. Hence, there exists a solution to the nonlinear equation (3.31).

Example 3.6.7 Consider the fractional damped dynamical system

$$\left. \begin{array}{l} \mathcal{D}^{7/4}x_1(t) - x_1(t) = \dfrac{\exp(-2t)\left(|x_1| + |\mathcal{D}^{1/2}x_1(t)|\right)}{1 + |x_2(t)|}, \\[4mm] \mathcal{D}^{7/4}x_2(t) - x_2(t) = \dfrac{\exp(-2t)\left(|x_2| + |\mathcal{D}^{1/2}x_2(t)|\right)}{1 + |x_1(t)|}, \end{array} \right\} \tag{3.33}$$

with initial conditions $\begin{bmatrix} x_1(0) \\ x_2(0) \end{bmatrix} = \begin{bmatrix} 1 \\ 0 \end{bmatrix}$ and $\begin{bmatrix} x_1'(0) \\ x_2'(0) \end{bmatrix} = \begin{bmatrix} 0 \\ 1 \end{bmatrix}$ for $t \in [0, 3]$.

It has the following form

$$\left. \begin{array}{l} \mathcal{D}^{7/4}x(t) + A^2 x(t) = f(t, x(t), {}^C D_{0+}^{1/2}x(t)), \ t \in [0, 3] \\ x(0) = x_0, \ x'(0) = y_0, \end{array} \right\} \tag{3.34}$$

where $A^2 = \begin{bmatrix} -1 & 0 \\ 0 & -1 \end{bmatrix}$, $x(t) = \begin{bmatrix} x_1(t) \\ x_2(t) \end{bmatrix}$, and

$$f(t, x(t), {}^C D_{0+}^{1/2}x(t)) = \begin{bmatrix} \dfrac{\exp(-2t)\left(|x_1| + |\mathcal{D}^{1/2}x_1(t)|\right)}{1 + |x_2(t)|} \\[4mm] \dfrac{\exp(-2t)\left(|x_2| + |\mathcal{D}^{1/2}x_2(t)|\right)}{1 + |x_1(t)|} \end{bmatrix}.$$

Using Mittag–Leffler matrix function for a given matrix A^2, we get

$$\Phi(3 - t) = \begin{bmatrix} N(t) & 0 \\ 0 & N(t) \end{bmatrix},$$

where $N(t) = (3 - t)^{3/4}E_{7/4,7/4}((3 - t)^{7/4})$. Further the nonlinear function f is continuous and satisfies the hypotheses of Theorem 3.5.1. Hence, the Eq. (3.33) has a solution on $[0, 3]$.

Example 3.6.8 Consider the following fractional differential equation

$$D^\alpha x(t) = f(t), \ t > 0 \tag{3.35}$$

where $0 < \alpha < 1$.

Take the initial condition $x(0) = 0$ and assume that the function $f(t)$ can be expressed as Taylor series

$$f(t) = \sum_{n=0}^{\infty} \frac{f^{(n)}(0)}{n!}t^n.$$

We know that

$$D^\alpha t^\beta = \frac{\Gamma(1+\beta)}{\Gamma(1+\beta-\alpha)} t^{\beta-\alpha}.$$

We can look for the solution of the Eq. (3.35) in the form of power series as

$$x(t) = t^\alpha \sum_{n=0}^{\infty} x_n t^n = \sum_{n=0}^{\infty} x_n t^{n+\alpha} \tag{3.36}$$

Substituting the expression (3.36) into the Eq. (3.35), we have

$$\sum_{n=0}^{\infty} x_n \frac{\Gamma(1+n+\alpha)}{\Gamma(n+1)} t^n = f(t) = \sum_{n=0}^{\infty} \frac{f^{(n)}(0)}{n!} t^n$$

and comparison of the coefficients of both sides gives

$$x_n = \sum_{n=0}^{\infty} \frac{f^{(n)}(0)}{\Gamma(1+n+\alpha)}, \quad n = 1, 2, \dots$$

Therefore, under the above assumptions, the solution of the Eq. (3.35) is

$$x(t) = t^\alpha \sum_{n=0}^{\infty} \frac{f^{(n)}(0)}{\Gamma(1+n+\alpha)} t^n. \tag{3.37}$$

In the case of the Eq. (3.35), we can easily transform the expression (3.37)

$$x(t) = \sum_{n=0}^{\infty} \frac{f^{(n)}(0)}{n!} \frac{\Gamma(n+1)}{\Gamma(1+n+\alpha)} t^{n+\alpha}$$

$$= \sum_{n=0}^{\infty} \frac{f^{(n)}(0)}{n!} I^\alpha t^n$$

$$= I^\alpha \left[\sum_{n=0}^{\infty} \frac{f^{(n)}(0)}{n!} t^n \right]$$

$$= I^\alpha f(t). \tag{3.38}$$

Applying I^α on both sides of (3.35) and the composition law of Riemann–Liouville derivative we would get the expression (3.38). However, the use of the inverse operator is often impossible.

3.7 Exercises

3.1. Solve $\sqrt{t}\, D^{\frac{1}{2}} x(t) + x(t) = 0$.

3.2. Solve $Dx(t) + D^{\frac{1}{2}} x(t) - 2x(t) = 0$.

3.3. Show that $DI^\alpha f(t) + I^\alpha Df(t) = \frac{f(0)}{\Gamma(\alpha)} t^{\alpha-1}$.

3.4. Solve ${}^C D^\alpha x(t) + \omega^\alpha x(t) = 0$, $1 < \alpha \le 2$, with $x(0) = x_0$ and $\dot{x}(0) = 0$.

3.5. Solve ${}^C D^\alpha x(t) + \kappa\, {}^C D^\beta x(t) + x(t) = 0$, $1 < \alpha \le 2$, $0 < \beta \le 1$, $\kappa > 0$, with $x(0) = 1$ and $\dot{x}(0) = 0$.

3.6. Solve ${}^C D^\alpha \theta(t) + \kappa \theta(t) = \kappa a$, $0 < \alpha \le 1$, with $\theta(0) = \theta_0$ and a, κ are constants.

3.7. Solve $(1 + t)\, {}^C D^\alpha y(t) = 1$ with $y(0) = 0$ and $0 < \alpha \le 1$.

3.8. Solve ${}^C D^\alpha y(t) = 1 - e^{-t}$, $0 < \alpha \le 1$ with $y(0) = 0$.

3.9. Solve ${}^C D^\alpha y(t) + {}^C D^\beta y(t) = 1$, $1 < \alpha \le 2$, $0 < \beta \le 1$, with $y(0) = y'(0) = 0$.

3.10. Solve the equation $D^2 y(t) + 2^C D^{\frac{3}{2}} y(t) + 2y(t) = \sin t$ with $y(0) = y'(0) = 0$.

References

1. Bonilla, B., Rivero, M., Rodriguez-Germá, L., Trujillo, J.J.: Fractional differential equations as alternative models to nonlinear differential equations. Appl. Math. Comput. **187**, 79–88 (2007)
2. Das, S.: Functional Fractional Calculus for Systems Identifications and Controls. Springer, New York (2008)
3. Diethelm, K.: The Analysis of Fractional Differential Equations. Springer, New York (2010)
4. Mainardi, F.: Fractional Calculus and Waves in Linear Viscoelasticity: An Introduction to Mathematical Models. Imperial College Press, London (2010)
5. Miller, K., Ross, B.: An Introduction to the Fractional Calculus and Fractional Differential Equations. Wiley, New York (1993)
6. Petráš, I.: Fractional-Order Nonlinear Systems: Modeling, Analysis and Simulation. Springer, Heidelberg (2011)
7. Podlubny, I.: Fractional Differential Equations. Academic Press, San Diego (1999)
8. Sabatier, J., Agrawal, O.P., Tenreiro Machado, J.A.: Advances in Fractional Calculus. Springer, Dordrecht (2007)
9. Samko, S.G., Kilbas, A.A., Marichev, O.I.: Fractional Integrals and Derivatives, Theory and Applications. Gordon and Breach Science Publishers, Amsterdam (1993)
10. Javidi, M., Nyamoradi, N.: Numerical behavior of a fractional order HIV/AIDS epidemic model. World J. Model. Simul. **9**, 139–149 (2013)
11. Rihan, F.A.: Numerical modeling of fractional-order biological systems. Abstract Appl. Anal. (2013), Art.ID 816803(11pp)
12. El-Sayed, A.M.A., Nour, H.M., Raslan, W.E., El-Shazly, E.S.: Fractional parallel RLC circuit. Alexandria J. Math. **3**, 11–23 (2012)
13. Kilbas, A.A., Srivastava, H.M., Trujillo, J.J.: Theory and Applications of Fractional Differential Equations. Elsevier, Amsterdam (2006)
14. Oldham, K., Spanier, J.: The Fractional Calculus. Academic Press, New York (1974)

Chapter 4
Applications

Abstract In this chapter, we discuss some applications of fractional differential equations in control theory. The basic problems in control theory such as observability, controllability, and stability are considered for fractional dynamical systems. Observability and controllability of linear systems are studied via Grammian matrix. Sufficient conditions for the controllability of nonlinear fractional dynamical systems are established by means of the fixed point theorem. Stability of linear and nonlinear systems is discussed. Examples are provided to illustrate the theory and few exercises are given.

Keywords Fractional dynamical systems · Observability · Controllability · Stability · Nonlinear systems

In this chapter, we discuss some applications of fractional differential equations in control theory. Basic facts about control theory can be found in the books [1, 2]. The basic problems in control theory such as observability, controllability, and stability are studied to the fractional dynamical systems in the references [3–13].

4.1 Observability

Observability is one of the fundamental concepts in control theory. The theory of observability is based on the possibility to deduce the initial state of the system from observing its input–output behavior. This means that, from the system's output, it is possible to determine the behavior of the entire system.

Consider the fractional order linear time-invariant system

$$^{C}D^{\alpha}x(t) = Ax(t), \ 0 < \alpha < 1, \ t \in J = [0.T], \tag{4.1}$$

with linear observation

$$y(t) = Hx(t), \tag{4.2}$$

where $x \in \mathbb{R}^n$, $y \in \mathbb{R}^m$, A is an $n \times n$ matrix and H is an $m \times n$ matrix.

© The Author(s), under exclusive license to Springer Nature Singapore Pte Ltd. 2023
K. Balachandran, *An Introduction to Fractional Differential Equations*, Industrial and Applied Mathematics, https://doi.org/10.1007/978-981-99-6080-4_4

Definition 4.1.1 The system (4.1) and (4.2) is *observable* on an interval J if

$$y(t) = Hx(t) = 0, \quad t \in J,$$

implies

$$x(t) = 0, \quad t \in J.$$

Theorem 4.1.2 *The linear system (4.1) and (4.2) is observable on J if and only if the observability Grammian matrix*

$$M = \int_0^T E_\alpha(A^* t^\alpha) H^* H E_\alpha(A t^\alpha) dt$$

is positive definite. Here, $*$ denotes the matrix transpose.

Proof The solution $x(t)$ of (4.1) corresponding to the initial condition $x(0) = x_0$ is given by

$$x(t) = E_\alpha(A t^\alpha) x_0$$

and we have, for $y(t) = Hx(t) = H E_\alpha(A t^\alpha) x_0$,

$$\|y\|^2 = \int_0^T y^*(t) y(t) dt$$

$$= x_0^* \int_0^T E_\alpha(A^* t^\alpha) H^* H E_\alpha(A t^\alpha) dt \, x_0$$

$$= x_0^* M x_0$$

is a quadratic form in x_0. Clearly, M is an $n \times n$ symmetric matrix. If M is positive definite, then $y = 0$ implies $x_0^* M x_0 = 0$. Therefore, $x_0 = 0$. Hence, (4.1)–(4.2) is observable on J. If M is not positive definite, then there is some $x_0 \neq 0$ such that $x_0^* M x_0 = 0$. Then $x(t) = E_\alpha(A t^\alpha) x_0 \neq 0$, for $t \in J$, but $\|y\|^2 = 0$, so $y = 0$ and we conclude that (4.1)–(4.2) is not observable on J. Hence, the proof. $\qquad\square$

Alternatively, the following result is proved in [14].

Lemma 4.1.3 *The fractional system (4.1) and (4.2) is observable on an arbitrary interval $[0, T]$ if and only if*

$$rank \begin{bmatrix} H \\ HA \\ \vdots \\ HA^{n-1} \end{bmatrix} = n.$$

If the linear system (4.1) and (4.2) is observable on an interval J, then $x(0) = x_0$, the initial state for the solution on that interval is reconstructed directly from the observation $y(t) = HE_\alpha(At^\alpha)x_0$.

Definition 4.1.4 The $n \times n$ matrix function $R(t)$ defined on J is a reconstruction kernel if and only if

$$\int_0^T R(t)HE_\alpha(At^\alpha)dt = I.$$

Theorem 4.1.5 *There exists a reconstruction kernel $R(t)$ on J if and only if the system (4.1) and (4.2) is observable on J.*

Proof If a reconstruction kernel exists and satisfies

$$\int_0^T R(t)y(t)dt = \int_0^T R(t)HE_\alpha(At^\alpha)dt\, x_0 = x_0$$

and $y(t) = 0$, then $x_0 = 0$. So $x(t) = 0$ and we conclude that the system (4.1) and (4.2) is observable on J. If, on the other hand, the system (4.1) and (4.2) is observable on J, then from Theorem 4.1.2

$$M = \int_0^T E_\alpha(A^*t^\alpha)H^*HE_\alpha(At^\alpha)dt > 0.$$

Let

$$R_0(t) = M^{-1}E_\alpha(A^*t^\alpha)H^*, \quad t \in J.$$

Then we have

$$\int_0^T R_0(t)HE_\alpha(At^\alpha)dt = M^{-1}\int_0^T E_\alpha(A^*t^\alpha)H^*HE_\alpha(At^\alpha)dt = I,$$

so that $R_0(t)$ is a reconstruction kernel on J. $\qquad \square$

4.2 Controllability of Linear Systems

The problem of controllability of dynamical systems is widely used in analysis and the design of control system. Any control system is said to be controllable if every state corresponding to this process can be affected or controlled at respective time by some control signals. The control problems involved in the description of fractional dynamical system are much more advanced.

Consider the linear dynamical system represented by the fractional differential equation of the form

$$^C D^\alpha x(t) = Ax(t) + Bu(t), \quad t \in J = [0, T], \tag{4.3}$$
$$x(0) = x_0,$$

with $0 < \alpha < 1$, $x \in R^n$, $u \in R^m$, A is a $n \times n$ matrix and B is a $n \times m$ matrix. The solution of the system (4.3) is

$$x(t) = E_\alpha(At^\alpha)x_0 + \int_0^t (t - s)^{\alpha-1} E_{\alpha,\alpha}(A(t - s)^\alpha) Bu(s)ds. \tag{4.4}$$

Definition 4.2.1 System (4.3) is said to be controllable on J if, for every $x_0, x_1 \in R^n$, there exists a control $u(t)$ such that the solution $x(t)$ of (4.3) satisfies the conditions $x(0) = x_0$ and $x(T) = x_1$.

Theorem 4.2.2 *The linear control system (4.3) is controllable on $[0, T]$ if and only if the controllability Grammian matrix*

$$W = \int_0^T [E_{\alpha,\alpha}(A(T - s)^\alpha)B][E_{\alpha,\alpha}(A(T - s)^\alpha)B]^* ds$$

is positive definite, for some $T > 0$.

Proof Since W is positive definite, it is non-singular and therefore its inverse is well-defined. Define the control function as

$$u(t) = (T - t)^{1-\alpha} B^* E_{\alpha,\alpha}(A^*(T - t)^\alpha) W^{-1}[x_1 - E_\alpha(AT^\alpha)x_0]. \tag{4.5}$$

Substituting $t = T$ in (4.4) and inserting (4.5), we have

$$\begin{aligned}
x(T) &= E_\alpha(AT^\alpha)x_0 + \int_0^T (T - s)^{\alpha-1} E_{\alpha,\alpha}(A(T - s)^\alpha) B(T - s)^{1-\alpha} \\
&\quad \times B^* E_{\alpha,\alpha}(A^*(T - s)^\alpha) W^{-1}[x_1 - E_\alpha(AT^\alpha)x_0]ds \\
&= E_\alpha(AT^\alpha)x_0 + WW^{-1}[x_1 - E_\alpha(AT^\alpha)x_0] \\
&= x_1.
\end{aligned}$$

Thus, (4.3) is controllable.

On the other hand, if it is not positive definite, there exists a nonzero y such that

$$y^* Wy = 0,$$

that is,

$$y^* \int_0^T E_{\alpha,\alpha}(A(T-s)^\alpha)BB^*E_{\alpha,\alpha}(A^*(T-s)^\alpha)y \ ds = 0$$

$$y^* E_{\alpha,\alpha}(A(T-s)^\alpha)B = 0,$$

on $[0, T]$. Let $x_0 = [E_\alpha(AT^\alpha)]^{-1}y$. By the assumption, there exists an input u such that it steers x_0 to the origin in the interval $[0, T]$. It follows that

$$x(T) = 0 = E_\alpha(AT^\alpha)x_0 + \int_0^T (T-s)^{\alpha-1}E_{\alpha,\alpha}(A(T-s)^\alpha)Bu(s)ds$$

$$0 = y + \int_0^T (T-s)^{\alpha-1}E_{\alpha,\alpha}(A(T-s)^\alpha)Bu(s)ds.$$

Then

$$0 = y^*y + \int_0^T (T-s)^{\alpha-1}y^*E_{\alpha,\alpha}(A(T-s)^\alpha)Bu(s)ds.$$

But the second term is zero leading to the conclusion $y^*y = 0$. This is a contradiction to $y \neq 0$. Thus, W is positive definite. Hence, the proof. □

Lemma 4.2.3 *[14] The fractional control system (4.3) is controllable if and only if*

$$rank\ [B, AB, \ldots, A^{n-1}B] = n.$$

Consider the linear fractional dynamical system represented by the fractional differential equation

$$\left.\begin{array}{l} {}^C D^\alpha x(t) + A^2 x(t) = Bu(t),\ t \in J, \\ x(0) = x_0,\ x'(0) = y_0, \end{array}\right\} \tag{4.6}$$

where $1 < \alpha \leq 2$, $x(t) \in \mathbb{R}^n$, $u(t) \in L_2(J; \mathbb{R}^m)$ and A, B are matrices of dimensions $n \times n$, $m \times n$, respectively. The solution of the system (4.6) is

$$x(t) = \Phi_0(t)x_0 + \Phi_1(t)y_0 + \int_0^t \Phi(t-s)Bu(s)ds. \tag{4.7}$$

where Φ, Φ_0, Φ_1 are already defined in Sect. 3.3 in terms of Mittag–Leffler functions.

Definition 4.2.4 The system (4.6) is said to be controllable on J if, for each vectors $x_0, y_0, x_1 \in \mathbb{R}^n$, there exists a control $u(t) \in L_2(J, \mathbb{R}^m)$ such that the corresponding solution of (4.6) together with $x(0) = x_0$ satisfies $x(T) = x_1$.

We note that our controllability definition is concerned only with steering the state vector but not the velocity vector y_0 in (4.6).

Lemma 4.2.5 *[2] Let f_i, for $i = 1, 2, \ldots, n$, be $1 \times p$ vector valued continuous functions defined on $[t_1, t_2]$. Let F be an $n \times p$ matrix with f_i as its ith row. Then $f_1, f_2, \ldots, f_n$ are linearly independent on $[t_1, t_2]$ if and only if the $n \times n$ constant matrix*

$$W(t_1, t_2) = \int_{t_1}^{t_2} F(t)F^*(t)dt$$

is positive definite.

Theorem 4.2.6 *The following statements regarding the linear system (4.6) are equivalent:*
(a) The linear system (4.6) is controllable on J.
(b) The rows of $\Phi(t)B$ are linearly independent.
(c) The controllability Grammian

$$W = \int_0^T \Phi(T-s)BB^*\Phi^*(T-s)ds \qquad (4.8)$$

is positive definite.

Proof First, we prove that $(a) \implies (b)$. Suppose that the system (4.6) is controllable, but the rows of $\Phi(t)B$ are linearly dependent on J. Then there exists a nonzero constant $n \times 1$ row vector y^* such that

$$y^* \Phi(t) B = 0, \text{ for every } t \in J. \qquad (4.9)$$

We choose $x(0) = x_0 = 0$, $x'(0) = y_0 = 0$. Therefore, the solution of (4.6) becomes

$$x(t) = \int_0^t \Phi(t-s)Bu(s)ds.$$

Since the system (4.6) is controllable on J, taking $x(T) = y$, we have

$$x(T) = y = \int_0^T \Phi(T-s)Bu(s)ds,$$

$$yy^* = \int_0^T y^*\Phi(T-s)Bu(s)ds.$$

From (4.9), $yy^* = 0$ and hence $y = 0$. Hence, it contradicts our assumption that y is nonzero. Now we prove that $(b) \implies (a)$. Suppose that the rows of $\Phi(t)B$ are linearly independent of J. Therefore, by Lemma (4.2.5), the $n \times n$ constant matrix

$$W = \int_0^T \Phi(T-s)BB^*\Phi^*(T-s)\,ds$$

is positive definite. Now we define the control function as

$$u(t) = B^* \Phi^*(T-t) W^{-1} [x_1 - \Phi_0(T)x_0 - \Phi_1(T)y_0]. \qquad (4.10)$$

Substituting (4.10) in (4.7), we have

$$\begin{aligned}
x(T) &= \Phi_0(T)x_0 + \Phi_1(T)y_0 + \int_0^T \Phi(T-s)BB^*\Phi^*(T-s)W^{-1} \\
&\quad \times [x_1 - \Phi_0(T)x_0 - \Phi_1(T)y_0] \, ds \\
&= \Phi_0(T)x_0 + \Phi_1(T)y_0 + WW^{-1}[x_1 - \Phi_0(T)x_0 - \Phi_1(T)y_0] \\
&= x_1.
\end{aligned}$$

Thus, the system (4.6) is controllable. The implications $(b) \implies (c)$ and $(c) \implies (b)$ follow directly from the Lemma (4.2.5). Hence, the result. $\qquad \square$

Remark 4.2.7 It should be mentioned that the linear fractional dynamical system (4.6) reduces to the second order dynamical system for $\alpha = 2$ and it is of the form [2]

$$\frac{d^2 x(t)}{dt^2} + A^2 x(t) = Bu(t), \quad t \in J,$$

with the same initial conditions $x(0) = x_0$ and $x'(0) = y_0$. Further taking $\alpha = 2$ in (4.7) and (4.8), one can easily derive the solution and the controllability Grammian for the above second order dynamical system as

$$x(t) = \cos(At)\, x_0 + A^{-1}\sin(At)\, y_0 + \int_0^t A^{-1}\sin(A(t-s))Bu(s)ds,$$

$$W = \int_0^T A^{-1}\sin(A(t-s))BB^*(A^{-1}\sin(A(t-s)))^*ds.$$

Moreover, this problem is also steering the states only but not the velocity vector in (4.6).

4.3 Controllability of Nonlinear Systems

Consider the nonlinear fractional dynamical system represented by the fractional differential equation of the form

$$^{C}D^{\alpha}x(t) = Ax(t) + Bu(t) + f(t, x(t), u(t)), \qquad (4.11)$$

$$x(0) = x_0,$$

where $0 < \alpha < 1$, $x \in R^n$, $u \in R^m$ and A is an $n \times n$ matrix, B is an $n \times m$ matrix and $f : J \times R^n \times R^m \rightarrow R^n$ is continuous. Let us introduce the following notation. Denote Q as the Banach space of continuous $R^n \times R^m$ valued functions defined on the interval J with the norm $\|(x, u)\| = \|x\| + \|u\|$, where $\|x\| = \sup\{|x(t)| : t \in J\}$ and $\|u\| = \sup\{|u(t)| : t \in J\}$. That is, $Q = C_n(J) \times C_m(J)$, where $C_n(J)$ is the Banach space of continuous R^n valued functions defined on the interval J with the sup norm.

For each $(z, v) \in Q$, consider the linear fractional dynamical system

$$
{}^CD^\alpha x(t) = Ax(t) + Bu(t) + f(t, z(t), v(t)),
$$
$$
x(0) = x_0.
$$

Then the solution is given by

$$
x(t) = E_\alpha(At^\alpha)x_0 + \int_0^t (t - s)^{\alpha-1} E_{\alpha,\alpha}(A(t - s)^\alpha) Bu(s)ds
$$
$$
+ \int_0^t (t - s)^{\alpha-1} E_{\alpha,\alpha}(A(t - s)^\alpha) f(s, z(s), v(s))ds. \qquad (4.12)
$$

Now we prove the following theorem.

Theorem 4.3.1 *If the nonlinear function f is continuous and uniformly bounded on J and if the linear system (4.3) is controllable, then the nonlinear system (4.11) is controllable on J.*

Proof Define the operator $P : Q \rightarrow Q$ by

$$
P(z, v) = (x, u)
$$

where

$$
u(t) = (T - t)^{1-\alpha} B^* E_{\alpha,\alpha}(A^*(T - t)^\alpha) W^{-1} \Bigg[x_1 - E_\alpha(AT^\alpha)x_0
$$
$$
- \int_0^T (T - s)^{\alpha-1} E_{\alpha,\alpha}(A(T - s)^\alpha) f(s, z(s), v(s))ds \Bigg] \qquad (4.13)
$$

and

$$
x(t) = E_\alpha(At^\alpha)x_0 + \int_0^t (t - s)^{\alpha-1} E_{\alpha,\alpha}(A(t - s)^\alpha) Bu(s)ds
$$
$$
+ \int_0^t (t - s)^{\alpha-1} E_{\alpha,\alpha}(A(t - s)^\alpha) f(s, z(s), v(s))ds. \qquad (4.14)
$$

Let

$$
a_1 = \sup \|E_{\alpha,\alpha}(A(T - t)^\alpha)\|, \quad a_2 = \sup \|E_\alpha(At^\alpha)x_0\|,
$$

$$c_1 = 4a_1^2 T^\alpha \|B^*\| \|W^{-1}\| \alpha^{-1}, \quad c_2 = 4a_1 T^\alpha \alpha^{-1},$$

$$d_1 = 4a_1 \|B^*\| \|W^{-1}\| [|x_1| + a_2], \quad d_2 = 4a_2,$$

$$M = \sup |f| = \sup\{|f(s, z(s), v(s))| : s \in J\}.$$

Then

$$|u(t)| \leq \|B^*\| a_1 \|W^{-1}\| \left[|x_1| + a_2 + T^\alpha a_1 \alpha^{-1} \sup |f| \right]$$

$$\leq \left[\frac{d_1}{4a} + \frac{c_1}{4a} M \right] = a$$

$$|x(t)| \leq a_2 + \frac{a_1 T^\alpha \|B\|}{4a\alpha} a + \frac{a_1 T^\alpha}{\alpha} M = b$$

Choose a positive constant r such that $a \leq r/2$ and $b \leq r/2$. Thus, if

$$Q(r) = \{(z, v) \in Q : \|z\| \leq \frac{r}{2} \text{ and } \|v\| \leq \frac{r}{2}\},$$

then P maps $Q(r)$ into itself. Since f is continuous, it implies that the operator is continuous and hence is completely continuous by the application of Arzela–Ascoli's theorem. Since $Q(r)$ is closed, bounded and convex, the Schauder fixed point theorem guarantees that P has a fixed point $(z, v) \in Q(r)$ such that

$$P(z, v) = (z, v) \equiv (x, u).$$

Hence, we have

$$x(t) = E_\alpha(At^\alpha)x_0 + \int_0^t (t - s)^{\alpha-1} E_{\alpha,\alpha}(A(t - s)^\alpha) Bu(s)ds$$

$$+ \int_0^t (t - s)^{\alpha-1} E_{\alpha,\alpha}(A(t - s)^\alpha) f(s, x(s), u(s))ds. \qquad (4.15)$$

Thus, $x(t)$ is the solution of the system (4.11) and it is easy to verify that $x(T) = x_1$. Hence, the system (4.11) is controllable on J. $\qquad \square$

4.4 Stability

The stability of a system relates to its response to inputs or disturbances. A system which remains in a constant state unless affected by an external action and which returns to a constant state when the external action is removed can be considered to be stable. The study of the stability of fractional system can be carried out by studying the solutions of the differential equations that characterize them. One notices that the analysis of stability in fractional order system is more complicated than in ordinary system. For instance, consider the following two systems with initial conditions $x(0) = 1$ for $0 < b < 1$,

$$\frac{d}{dt}x(t) = bt^{b-1} \tag{4.16}$$

$$^{C}D^{\alpha}x(t) = bt^{b-1}, 0 < \alpha < 1. \tag{4.17}$$

The analytical solution of (4.16) is

$$x(t) = 1 + t^b.$$

As $t \to \infty$, $x(t) \to \infty$. Therefore, the integer-order system (4.16) is unstable for any $b \in (0, 1)$. The analytical solution of (4.17) is

$$x(t) = 1 + \frac{\Gamma(b+1)}{\Gamma(b+\alpha)}t^{b+\alpha-1}.$$

As $t \to \infty$, $x(t) \to 1$, when $b + \alpha - 1 < 0$. Therefore, the fractional order system (4.17) is stable for any $0 < b \leq 1 - \alpha$ and this implies that the fractional order system may have additional attractive feature over the integer-order system.

Consider the fractional linear time-invariant system

$$^{C}D^{\alpha}x(t) = Ax(t), \ 0 < \alpha < 1, \ t \in [t_0, t_1], \tag{4.18}$$
$$x(t_0) = x_0,$$

where $x \in R^n$ and A is an $n \times n$ matrix. The solution of (4.18) is given by

$$x(t) = E_{\alpha}(A(t - t_0)^{\alpha})x_0.$$

Definition 4.4.1 The solution of fractional dynamical system (4.18) is said to be stable if there exists $\varepsilon > 0$ such that any solution $x(t)$ of (4.18) satisfies $\|x(t)\| < \varepsilon$ for all $t > t_0$. The solution is said to be asymptotically stable if in addition to being stable, $\|x(t)\| \to 0$ as $t \to \infty$.

Theorem 4.4.2 *The fractional dynamical system (4.18) is asymptotically stable iff*

$$|\arg(spec A)| > \alpha\pi/2.$$

Proof Taking Laplace transform on both sides of (4.18), we get

$$X(s)s^\alpha - s^{\alpha-1}x_0 = AX(s).$$

It follows that the solution of the linear system (4.18) is given by

$$x(t) = E_\alpha(At^\alpha)x_0.$$

First, suppose that the matrix A is similar to the diagonal matrix. Then there exists an invertible matrix T such that

$$A = T^{-1}AT = diag(\lambda_1, \ldots, \lambda_n).$$

Then

$$E_\alpha(At^\alpha) = TE_\alpha(At^\alpha)T^{-1} = Tdiag[E_\alpha(\lambda_1 t^\alpha), E_\alpha(\lambda_2 t^\alpha), \ldots, E_\alpha(\lambda_n t^\alpha)]T^{-1}.$$

By a result in [15], we have

$$E_\alpha(\lambda_i t^\alpha) = -\sum_{k=2}^{p} \frac{(\lambda_i t^\alpha)^{-k}}{\Gamma(1 - k\alpha)} + O(|\lambda_i t^\alpha|^{-1-p}) \to 0 \text{ as } t \to +\infty, \ 1 \le i \le n.$$

Hence, the conclusion holds. Next, suppose the matrix A is similar to a Jordan canonical form, that is, there exists an invertible matrix T such that $J = T^{-1}AT = diag(J_1, \ldots, J_r)$, where J_i, $1 \le i \le r$ has the following form:

$$J_i = \begin{pmatrix} \lambda_i & 1 & & \\ & \lambda_i & \ddots & \\ & & \ddots & 1 \\ & & & \lambda_i \end{pmatrix}_{n_i \times n_i}$$

and $\sum_{i=1}^{r} n_i = n$. Obviously,

$$E_\alpha(At^\alpha) = Tdiag[E_\alpha(J_1 t^\alpha), E_\alpha(\lambda_2 t^\alpha), \ldots, E_\alpha(J_r t^\alpha)]T^{-1},$$

where for $1 \le i \le r$,

$$E_\alpha(J_i t^\alpha) = \sum_{k=0}^{\infty} \frac{(J_i t^\alpha)^k}{\Gamma(\alpha k + 1)} = \sum_{k=0}^{\infty} \frac{(t^\alpha)^k}{\Gamma(\alpha k + 1)} J_i^k$$

$$= \sum_{k=0}^{\infty} \frac{(t^\alpha)^k}{\Gamma(\alpha k + 1)} \begin{pmatrix} \lambda_i^k & C_k^1 \lambda_i^{k-1} & \cdots & C_k^{n_i-1} \lambda_i^{k-n_i+1} \\ & \lambda_i^k & \ddots & \vdots \\ & & \ddots & C_k^1 \lambda_i^{k-1} \\ & & & \lambda_i^k \end{pmatrix}$$

$$= \begin{pmatrix} \sum_{k=0}^{\infty} \frac{(\lambda_i t^\alpha)^k}{\Gamma(\alpha k+1)} & \sum_{k=0}^{\infty} \frac{(t^\alpha)^k}{\Gamma(\alpha k+1)} C_k^1 \lambda_i^{k-1} & \cdots & \sum_{k=0}^{\infty} \frac{(t^\alpha)^k}{\Gamma(\alpha k+1)} C_k^{n_i-1} \lambda_i^{k-n_i+1} \\ & \sum_{k=0}^{\infty} \frac{(\lambda_i t^\alpha)^k}{\Gamma(\alpha k+1)} & \ddots & \vdots \\ & & \ddots & \sum_{k=0}^{\infty} \frac{(t^\alpha)^k}{\Gamma(\alpha k+1)} C_k^1 \lambda_i^{k-1} \\ & & & \sum_{k=0}^{\infty} \frac{(\lambda_i t^\alpha)^k}{\Gamma(\alpha k+1)} \end{pmatrix}$$

($C_k^j, 1 \le j \le n_i - 1$ are the binomial coefficients)

$$= \begin{pmatrix} E_\alpha(\lambda_i t^\alpha) & \frac{1}{(1)!}\left(\frac{\partial}{\partial \lambda_i}\right) E_\alpha(\lambda_i t^\alpha) & \cdots & \frac{1}{(n_i-1)!}\left(\frac{\partial}{\partial \lambda_i}\right)^{n_i-1} E_\alpha(\lambda_i t^\alpha) \\ & E_\alpha(\lambda_i t^\alpha) & \ddots & \vdots \\ & & \ddots & \frac{1}{(1)!}\left(\frac{\partial}{\partial \lambda_i}\right) E_\alpha(\lambda_i t^\alpha) \\ & & & E_\alpha(\lambda_i t^\alpha) \end{pmatrix}.$$

By some tedious calculations, if $|arg(\lambda_i(A))| > \frac{\alpha\pi}{2}, 1 \le i \le r$ and $t \to \infty$, then we have

$$E_\alpha(\lambda_i t^\alpha) \to 0 \quad \text{and} \quad \left| \frac{1}{j!}\left(\frac{\partial}{\partial \lambda_i}\right)^j E_\alpha(\lambda_i t^\alpha) \right| \to 0, \quad 0 \le j \le n_i - 1, \quad 1 \le i \le r.$$

Indeed, these can be seen from the following:

$$E_\alpha(\lambda_i t^\alpha) = -\sum_{k=2}^{p} \frac{(\lambda_i t^\alpha)^{-k}}{\Gamma(1 - k\alpha)} + O(|\lambda_i t^\alpha|^{-1-p}),$$

which implies $E_\alpha(\lambda_i t^\alpha) \to 0$ as $t \to \infty$ and

$$\frac{1}{j!}\left(\frac{\partial}{\partial \lambda_i}\right)^j E_\alpha(\lambda_i t^\alpha) = \frac{1}{j!}\left(\frac{\partial}{\partial \lambda_i}\right)^j \left\{ -\sum_{k=2}^{p} \frac{(\lambda_i t^\alpha)^{-k}}{\Gamma(1 - k\alpha)} + O(|\lambda_i t^\alpha|^{-1-p}) \right\}$$

$$= -\sum_{k=2}^{p} \frac{(-1)^j (k+j-1)\ldots(k+1)k \lambda_i^{-k-j} t^{-\alpha k}}{j! \Gamma(1 - \alpha k)}$$

$$\qquad + O(|\lambda_i|^{-1-p-j} |t^\alpha|^{-1-p})$$

$$= -\sum_{k=2}^{p} \frac{(-1)^j (k+j-1)! \lambda_i^{-k-j} t^{-\alpha k}}{j!(k - 1)! \Gamma(1 - \alpha k)} + O(|\lambda_i|^{-1-p-j} |t^\alpha|^{-1-p})$$

which leads to $\left|\frac{1}{j!}\left(\frac{\partial}{\partial\lambda_i}\right)^j E_\alpha(\lambda_i t^\alpha)\right| \to 0, \quad 0 \le j \le n_i - 1$ as $t \to \infty$. It now follows that

$$\|x(t)\| = \|E_\alpha(At^\alpha)x_0\| \to 0 \text{ as } t \to \infty$$

for any nonzero initial value x_0. The proof is complete. □

The following results are useful in analyzing the stability results of fractional differential equations.

Lemma 4.4.3 *[16] If all the eigenvalues of A satisfy*

$$|arg(spec(A))| > \frac{\alpha\pi}{2}, \tag{4.19}$$

then the solution of the system (4.18) is asymptotically stable.

Lemma 4.4.4 *[15] If $\alpha < 2$ and β is arbitrary real number, μ is such that $\alpha\pi/2 < \mu < min\{\pi, \alpha\pi\}$ and C is a real constant, then*

$$|E_{\alpha,\beta}(z)| \le \frac{C}{1 + |z|}.$$

$(\mu \le |arg(z)| \le \pi, |z| \ge 0)$

Lemma 4.4.5 (Gronwall Lemma) *[17] Suppose that g(t) and f(t) are non-negative continuous functions on $[t_0, t_1]$, $\lambda \ge 0$ and if*

$$f(t) \le \lambda + \int_{t_0}^t g(s)f(s)ds,$$

then

$$f(t) \le \lambda e^{\int_{t_0}^t g(s)ds}, \quad t_0 \le t \le t_1.$$

Theorem 4.4.6 *Suppose $\|E_{\alpha,\beta}(At^\alpha)\| \le Me^{-\gamma t}, 0 \le t < \infty, \gamma > 0$ for $\beta = 1, \beta = \alpha$ and $\int_0^\infty \|B(t)\|dt < N$, where $M, N > 0$, then the solution of equation*

$$^C D^\alpha x(t) = Ax(t) + B(t)x(t), \quad 0 < \alpha < 1, \tag{4.20}$$
$$x(0) = x_0,$$

is asymptotically stable.

Proof By the Laplace transform and the inverse Laplace transform the solution of Eq. (4.20) can be written as

$$x(t) = E_\alpha(At^\alpha)x_0 + \int_0^t (t - \tau)^{\alpha-1} E_{\alpha,\alpha}(A(t - \tau)^\alpha) B(\tau) x(\tau) d\tau.$$

Then we obtain

$$\|x(t)\| \leq \|E_\alpha(At^\alpha)\| \|x_0\| + \int_0^t \|(t - \tau)^{\alpha-1} E_{\alpha,\alpha}(A(t - \tau)^\alpha)\| \|B(\tau)\| \|x(\tau)\| d\tau,$$

from the boundedness, we can obtain

$$\|x(t)\| \leq M e^{-\gamma t} \|x_0\| + \int_0^t M e^{-\gamma(t-\tau)} \|B(\tau)\| \|x(\tau)\| d\tau. \qquad (4.21)$$

Multiplying by $e^{\gamma t}$ both sides of Eq. (4.21), we have

$$e^{\gamma t} \|x(t)\| \leq M \|x_0\| + \int_0^t M e^{\gamma\tau} \|B(\tau)\| \|x(\tau)\| d\tau.$$

Let $e^{\gamma t} \|x(t)\| = u(t)$. Then according to the previous lemma, we have

$$e^{\gamma t} \|x(t)\| \leq M \|x_0\| \exp\left(\int_0^t \|B(s)\| ds\right). \qquad (4.22)$$

Multiplying by $e^{-\gamma t}$ both sides of Eq. (4.22), we obtain

$$\|x(t)\| \leq M \|x_0\| \exp\left(\int_0^t \|B(s)\| ds\right) e^{-\gamma t}.$$

Then $\|x(t)\| \leq M \|x_0\| e^{N-\gamma t}$, so $\|x(t)\| \to 0$, $t \to \infty$, that is, the solution of Eq. (4.20) is asymptotically stable. $\square$

4.5 Nonlinear Equations

Consider the nonlinear fractional differential equation of order $0 < \alpha < 1$

$$^C D^\alpha x(t) = f(t, x(t)), \ t \in J = [0, T], \qquad (4.23)$$
$$x(0) = x_0,$$

where $f : J \times \mathbb{R}^n \to \mathbb{R}^n$ is continuous and $f(t, 0) = 0$.

Definition 4.5.1 The zero solution of the Eq. (4.23) is said to be stable if, for any initial value x_0, there exists $\epsilon > 0$ such that $\|x(t)\| < \epsilon$ for all $t > 0$. The zero solution is said to be asymptotically stable if, in addition to being stable, $\|x(t)\| \to 0$ as $t \to +\infty$.

Consider the nonlinear fractional system of the form

$$^C D^\alpha x(t) = Ax(t) + f(t, x(t)), \tag{4.24}$$
$$x(0) = x_0,$$

where $0 < \alpha < 1$, A is an $n \times n$ matrix and $f(t, x(t)) \in \mathbb{C}(J \times \mathbb{R}^n, \mathbb{R}^n)$ with $f(t, 0) = 0$.

Theorem 4.5.2 *Suppose that* $\|f(t, x(t))\| \le M\|x\|$ *and all the eigenvalues of A satisfy (4.19). Then the zero solution of (4.24) is asymptotically stable.*

Proof The solution of the system (4.24) can be written as

$$x(t) = E_\alpha(At^\alpha)x_0 + \int_0^t (t - s)^{\alpha-1} E_{\alpha,\alpha}(A(t - s)^\alpha) f(s, x(s)) ds.$$

From this,

$$\|x(t)\| \le \|E_\alpha(At^\alpha)x_0\| + \int_0^t \|(t - s)^{\alpha-1} E_{\alpha,\alpha}(A(t - s)^\alpha)\| \| f(s, x(s))\| ds$$

$$\le \|E_\alpha(At^\alpha)x_0\| + M \int_0^t \|(t - s)^{\alpha-1} E_{\alpha,\alpha}(A(t - s)^\alpha)\| \|x\| ds.$$

By using the Gronwall Lemma, we have

$$\|x(t)\| \le \|E_\alpha(At^\alpha)x_0\| \exp\left\{ M \int_0^t \|(t - s)^{\alpha-1} E_{\alpha,\alpha}(A(t - s)^\alpha)\| ds \right\},$$

and so

$$\|x(t)\| \le C\|E_\alpha(At^\alpha)x_0\|,$$

where

$$C = \max_{t \in J} \exp\left\{ M \int_0^t \|(t - s)^{\alpha-1} E_{\alpha,\alpha}(A(t - s)^\alpha)\| ds \right\}.$$

Further, $\|E_\alpha(At^\alpha)x_0\| \to 0$ as $t \to \infty$. Hence, we have

$$\lim_{t \to \infty} x(t) = 0.$$

Therefore, the zero solution of the given system is asymptotically stable. □

Consider the nonlinear system of the form

$$^C D^\alpha x(t) = Ax(t) + I^\alpha g(t, x(t)), \quad 0 < \alpha \le 1, \tag{4.25}$$
$$x(0) = x_0$$

where A is an $n \times n$ matrix and $g(t, x(t)) \in \mathbb{C}[J \times \mathbb{R}^n, \mathbb{R}^n]$, $g(t, 0) = 0$.

Lemma 4.5.3 *[18] If A satisfies (4.19), then there exists a constant $K > 0$ such that*

$$\int_0^t ||t^{2\alpha-1} E_{\alpha,2\alpha}(At^\alpha)||dt \leq K. \tag{4.26}$$

Theorem 4.5.4 *Let $g(t, x(t))$ satisfy*

$$||g(t, x(t))|| \leq M||x||, \tag{4.27}$$

and all the eigenvalues of A satisfy (4.19). Then the zero solution of (4.25) is asymptotically stable.

Proof The solution of the system can be written as

$$
\begin{aligned}
x(t) &= E_\alpha(At^\alpha)x_0 + \int_0^t (t-s)^{\alpha-1} E_{\alpha,\alpha}(A(t-s)^\alpha)I^\alpha g(s, x(s))ds \\
&= E_\alpha(At^\alpha)x_0 + \frac{1}{\Gamma(\alpha)} \int_0^t (t-s)^{\alpha-1} E_{\alpha,\alpha}(A(t-s)^\alpha) \int_0^s (s-\tau)^\alpha g(\tau, x(\tau))d\tau ds \\
&= I_1 + I_2.
\end{aligned}
$$

Let us evaluate I_2. For,

$$
\begin{aligned}
I_2 &= \frac{1}{\Gamma(\alpha)} \int_0^t \int_0^s (t-s)^{\alpha-1}(s-\tau)^{\alpha-1} E_{\alpha,\alpha}(A(t-s)^\alpha)g(\tau, x(\tau))d\tau ds \\
&= \frac{1}{\Gamma(\alpha)} \int_0^t g(\tau, x(\tau)) \sum_{k=0}^\infty \frac{A^k}{\Gamma(\alpha k + \alpha)} \int_\tau^t (t-s)^{\alpha k+\alpha-1}(s-\tau)^{\alpha-1}ds d\tau \\
&= \int_0^t (t-\tau)^{2\alpha-1} \sum_{k=0}^\infty \frac{(A(t-\tau)^\alpha)^k}{\Gamma(\alpha k + 2\alpha)}g(\tau, x(\tau))d\tau \\
&= \int_0^t (t-\tau)^{2\alpha-1} E_{\alpha,2\alpha}(A(t-\tau)^\alpha)g(\tau, x(\tau))d\tau.
\end{aligned}
$$

Hence

$$x(t) = E_\alpha(At^\alpha)x_0 + \int_0^t (t-\tau)^{2\alpha-1} E_{\alpha,2\alpha}(A(t-\tau)^\alpha)g(\tau, x(\tau))d\tau.$$

Then

$$||x(t)|| \leq ||E_\alpha(At^\alpha)x_0|| + M \int_0^t (t-\tau)^{2\alpha-1} E_{\alpha,2\alpha}(A(t-\tau)^\alpha)||x||d\tau.$$

By using Gronwall's Lemma,

$$\|x(t)\| \leq \|E_\alpha(At^\alpha)x_0\| \exp\left\{M \int_0^t \|(t-\tau)^{2\alpha-1} E_{\alpha,2\alpha}(A(t-\tau)^\alpha)\| d\tau\right\}$$

and so

$$\|x(t)\| \leq C \|E_\alpha(At^\alpha)x_0\|,$$

where

$$C = \max_{t\in J} \exp\left\{M \int_0^t \|(t-\tau)^{2\alpha-1} E_{\alpha,2\alpha}(A(t-\tau)^\alpha)\| d\tau\right\}.$$

Further, $\|E_\alpha(At^\alpha)x_0\| \to 0$ as $t \to \infty$. Hence, we have $\lim_{t\to\infty} x(t) = 0$. Therefore the zero solution of the given system is asymptotically stable. $\qquad\square$

Consider the fractional integrodifferential equation of the form

$$^C D^\alpha x(t) = Ax(t) + g\left(t, x(t), \int_0^t k(t, s, x(s))ds\right), \tag{4.28}$$

with the initial condition $x(0) = x_0$, where $0 < \alpha \leq 1$, $g \in C[J \times \mathbb{R}^n \times \mathbb{R}^n, \mathbb{R}^n]$ and $k \in C[J \times J \times \mathbb{R}^n, \mathbb{R}^n]$ with $g(t, 0, 0) = 0$, $k(t, s, 0) = 0$ for all $t, s \in J$.

Theorem 4.5.5 *Let $k(t, s, x(s))$ satisfy*

$$\|k(t, s, x(s))\| \leq M_1 \|x\|, \quad s \in [0, t], \tag{4.29}$$

and all the eigenvalues of A satisfy (4.19). Then the zero solution of (4.28) is asymptotically stable.

Proof Comparing the Eq. (4.28) with the Eq. (4.24), the nonlinear term is given by

$$f(t, x(t)) = g\left(t, x(t), \int_0^t k(t, s, x(s))ds\right).$$

Then the condition for the stability is given by

$$\|f(t, x(t))\| = \left\|g\left(t, x(t), \int_0^t k(t, s, x(s))ds\right)\right\|$$

$$\leq C_1 \|x\| + C_2 \left\|\int_0^t k(t, s, x(s))ds\right\|.$$

By using the condition (4.29), we have

$$\|g(t, x(t), \int_0^t k(t, s, x(s))\mathrm{d}s)\| \leq C_1\|x\| + C_2 T M_1\|x\|$$
$$\leq M\|x\|,$$

where $M = C_1 + C_2 T M_1$. Hence, the nonlinear term satisfies the required condition of the Theorem 4.5.2 and the rest of the proof is similar to that of Theorem 4.5.2. Thus the system (4.28) is asymptotically stable.   □

Corollary 4.5.6 *Consider the nonlinear system of the form*

$$^C D^\alpha x(t) = Ax(t) + \int_0^t g(\tau, x(\tau))\mathrm{d}\tau, \ \alpha \in (0, 1), \tag{4.30}$$

where A is an $n \times n$ matrix and $g(t, x) \in \mathbb{C}(J \times \mathbb{R}^n, \mathbb{R}^n)$, $g(t, 0) = 0$ with the initial condition given by $x(0) = 0$. Suppose that

$$\|g(t, x(t))\| \leq M_1\|x\| \tag{4.31}$$

and all the eigenvalues of A satisfy (4.19). Then the zero solution of (4.30) is asymptotically stable.

Consider the nonlinear system of the form

$$^C D^\alpha x(t) - AI^\alpha x(t) = f(t, x(t)), \ \alpha \in (0, 1), \tag{4.32}$$

where A is an $n \times n$ matrix and the initial condition is given by $x(0) = x_0$ and $f(t, 0) = 0$.

Theorem 4.5.7 *Suppose $f(t, x(t))$ satisfies the condition*

$$\|f(t, x(t))\| \leq M_1\|x\| \tag{4.33}$$

and all the eigenvalues of A satisfy (4.19). Then the zero solution of (4.32) is asymptotically stable.

Proof The given Eq. (4.32) can be written as

$$^C D^\alpha x(t) = AI^\alpha x(t) + f(t, x(t)).$$

Taking Laplace transform on both sides, we get

$$X(s) = \frac{s^{2\alpha} - 1}{s^{2\alpha} - A} x_0 + \mathcal{L}\{t^{2\alpha-1} E_{2\alpha,\alpha}(At^{2\alpha}) * f(t, x(t))\}.$$

By taking inverse Laplace transform, the solution representation of the system can be written as

$$x(t) = E_{2\alpha}(At^{2\alpha})x_0 + \int_0^t (t-s)^{\alpha-1} E_{2\alpha,\alpha}(A(t-s)^{2\alpha}) f(s, x(s)) ds$$

and from this, we have

$$\|x(t)\| \leq \|E_{2\alpha}(At^{2\alpha})x_0\| + \int_0^t \|(t-s)^{\alpha-1} E_{2\alpha,\alpha}(A(t-s)^{2\alpha})\| \|f(s, x(s))\| ds.$$

By using Gronwall's inequality,

$$\|x(t)\| \leq \|E_{2\alpha}(At^{2\alpha})x_0\| \exp\left\{ M \int_0^t s^{\alpha-1} E_{2\alpha,\alpha}(A(s)^{2\alpha}) ds \right\}.$$

Since

$$\exp\left\{ M \int_0^t s^{\alpha-1} E_{2\alpha,\alpha}(As^{2\alpha}) \right\} ds$$

is bounded and $\|E_{2\alpha}(At^{2\alpha})x_0\| \to 0$ as $t \to \infty$, we have $\lim_{t\to\infty} x(t) = 0$.   □

4.6 Examples

In this section, we apply the results established in the previous sections to the following fractional dynamical systems. Let $x_1(t) = x(t)$ and $x_2(t) = {}^C D^\alpha x_1(t)$.

Example 4.6.1 Consider the linear fractional dynamical system

$$^C D^\alpha x(t) = Ax(t), \quad 0 < \alpha < 1, \quad t \in [0, T], \tag{4.34}$$
$$x(0) = x_0,$$

with the linear observation $y(t) = Hx(t)$ where $A = \begin{bmatrix} 0 & 1 \\ -1 & 0 \end{bmatrix}$, $H = \begin{bmatrix} 1 & 0 \end{bmatrix}$ and $x(t) = \begin{bmatrix} x_1(t) \\ x_2(t) \end{bmatrix}$. The Mittag–Leffler matrix function is given by

$$E_\alpha(At^\alpha) = \begin{bmatrix} \frac{1}{2}[E_\alpha(it^\alpha) + E_\alpha(-it^\alpha)] & \frac{1}{2i}[E_\alpha(it^\alpha) - E_\alpha(-it^\alpha)] \\ -\frac{1}{2i}[E_\alpha(it^\alpha) - E_\alpha(-it^\alpha)] & \frac{1}{2}[E_\alpha(it^\alpha) + E_\alpha(-it^\alpha)] \end{bmatrix}.$$

The observability Grammian for this system is

$$M = \int_0^T E_\alpha(A^* t^\alpha) H^* H E_\alpha(A t^\alpha) dt.$$

Using matrix calculations, we can show that M is positive definite for $T > 0$. Hence, the system (4.34) is observable.

Example 4.6.2 Consider the following nonlinear fractional dynamical system represented by the scalar fractional differential equation

$$^C D^{1/2} x(t) = x(t) + u(t) + \sin x(t) \cos u(t), \quad t \in [0, 1], \qquad (4.35)$$
$$x(0) = x_0,$$

here $A = B = 1$; $\alpha = 1/2$; $T = 1$; $f(t, x(t), u(t)) = \sin x(t) \cos u(t)$.

The two-parameter Mittag–Leffler function is given by

$$E_{1/2, 1/2}((t - s)^{1/2}) = \sum_{k=0}^{\infty} \frac{(t - s)^{k/2}}{\Gamma((k + 1)/2)}.$$

By simple calculation, one can see that the controllability Grammian

$$
\begin{aligned}
W &= \int_0^1 [E_{1/2, 1/2}((1 - s)^{1/2})][E_{1/2, 1/2}((1 - s)^{1/2})]^* ds \\
&= \int_0^1 \sum_{k=0}^{\infty} \frac{(1 - s)^{k/2}}{\Gamma((k + 1)/2)} \times \sum_{m=0}^{\infty} \frac{(1 - s)^{m/2}}{\Gamma((m + 1)/2)} ds \\
&= \int_0^1 \sum_{k=0}^{\infty} \sum_{m=0}^{\infty} \frac{(1 - s)^{(k+m)/2}}{\Gamma((k + 1)/2)\Gamma((m + 1)/2)} ds \\
&= \sum_{k=0}^{\infty} \sum_{m=0}^{\infty} \frac{2}{(k + m + 2)\Gamma((k + 1)/2)\Gamma((m + 1)/2)} \\
&> 0
\end{aligned}
$$

and the control function is

$$
\begin{aligned}
u(t) = \sum_{k=0}^{\infty} \frac{(1 - t)^{(k+1)/2}}{\Gamma((k + 1)/2)} W^{-1} \Bigg[&x_1 - \sum_{k=0}^{\infty} \frac{(1)^{k/2}}{\Gamma((k/2 + 1)} x_0 \\
&- \sum_{k=0}^{\infty} \int_0^1 \frac{(1 - s)^{(k-1)/2}}{\Gamma((k + 1)/2)} \sin x(s) \cos u(s) ds \Bigg];
\end{aligned}
$$

since $W > 0$, the linear system is controllable and the nonlinear function $f(t, x, u) = \sin x \cos u$ is uniformly bounded. Hence, by Theorem 4.3.1, the nonlinear system (4.35) is controllable on $[0, 1]$.

Example 4.6.3 Consider the linear fractional control system

$$^{C}D^{\alpha}x(t) = Ax(t) + Bu(t), \quad 0 < \alpha < 1, \quad t \in [0, T], \quad (4.36)$$

where $A = \begin{bmatrix} 0 & 1 \\ 1 & 0 \end{bmatrix}$ and $B = \begin{bmatrix} 0 \\ 1 \end{bmatrix}$.

The Mittag–Leffler matrix function of the given matrix A is

$$E_{\alpha,\alpha}(At^{\alpha}) = \begin{bmatrix} E_{2\alpha,\alpha}(t^{2\alpha}) & t^{\alpha}E_{2\alpha,2\alpha}(t^{2\alpha}) \\ t^{\alpha}E_{2\alpha,2\alpha}(t^{2\alpha}) & E_{2\alpha,\alpha}(t^{2\alpha}) \end{bmatrix}.$$

The controllability Grammian of this system is

$$W = \int_{0}^{T} E_{\alpha,\alpha}(A(T - \tau)^{\alpha})BB^{*}E_{\alpha,\alpha}(A^{*}(T - \tau)^{\alpha})d\tau.$$

Using matrix calculations, we can show that W is positive definite for $T > 0$. So the linear system (4.36) is controllable on $[0, T]$.

Example 4.6.4 Consider the nonlinear fractional control system

$$^{C}D^{\alpha}x(t) = Ax(t) + Bu(t) + f(t, x, u), \quad 0 < \alpha < 1, \quad t \in [0, T], \quad (4.37)$$

where A, B are as above and $f(t, x, u) = \begin{pmatrix} \sin x_1 \\ \sin u \end{pmatrix}$.

Since the controllability Grammian W is positive definite for $T > 0$ and so the linear part of the system is controllable on $[0, T]$. Further, the nonlinear function f satisfies the hypothesis of the Theorem 4.3.1. Observe that the control defined by

$$u(t) = (T - t)^{1-\alpha}B^{*}E_{\alpha,\alpha}(A^{*}(T - t)^{\alpha})W^{-1}\left[x_1 - E_{\alpha}(A(T)^{\alpha})x_0\right.$$
$$\left. - \int_{0}^{T}(T - s)^{\alpha-1}E_{\alpha,\alpha}(A^{*}(T - t)^{\alpha})f(s, x(s), u(s))ds\right]$$

steers the system (4.37) from x_0 to x_1 and hence the nonlinear fractional control system is controllable on $[0, T]$.

Example 4.6.5 Consider the linear fractional dynamical system

$$^C D^\alpha x(t) = \begin{bmatrix} 0 & -1 \\ 1 & 0 \end{bmatrix} x(t), \ 0 < \alpha < 1, \ t \in [0, T], \tag{4.38}$$

$$x(0) = x_0,$$

The eigenvalues of the given matrix are $\pm i$.

$$|\arg(i)| = \left|\tan^{-1}\left(\frac{1}{0}\right)\right| = \left|\frac{\pi}{2}\right| = \frac{\pi}{2} > \frac{\alpha\pi}{2}, \quad 0 < \alpha < 1,$$

$$|\arg(-i)| = \left|\tan^{-1}\left(\frac{-1}{0}\right)\right| = \left|-\frac{\pi}{2}\right| = \frac{\pi}{2} > \frac{\alpha\pi}{2}, \quad 0 < \alpha < 1.$$

Therefore, the fractional order system (4.38) is asymptotically stable. This implies that the system is stable. When $\alpha = 1$, the integer order system is stable but not asymptotically stable.

Example 4.6.6 Consider the fractional dynamical system

$$^C D^\alpha x(t) = \begin{bmatrix} 1 & -1 \\ 1 & 1 \end{bmatrix} x(t), \ 0 < \alpha < 1, \ t \in [0, T], \tag{4.39}$$

$$x(0) = x_0,$$

The eigenvalues of the given matrix are $1 \pm i$.

$$|\arg(1+i)| = \left|\tan^{-1}(1)\right| = \left|\frac{\pi}{4}\right| = \frac{\pi}{4} > \frac{\alpha\pi}{2}, \text{ if } \alpha = 1/3$$

$$|\arg(1-i)| = \left|\tan^{-1}(-1)\right| = \left|\frac{\pi}{4}\right| = \frac{\pi}{4} > \frac{\alpha\pi}{2}, \text{ if } \alpha = 1/3.$$

Therefore, the fractional order system (4.39) is asymptotically stable and this implies that the system is stable if $\alpha = 1/3$. But, for $\alpha = 1$, the integer order system is unstable.

Example 4.6.7 Consider the following system of fractional differential equations in the Caputo sense with $\alpha \in (0, 1)$:

$$\begin{cases} ^C D^\alpha x = ax - by \\ ^C D^\alpha y = bx + ay \end{cases} \tag{4.40}$$

Applying the Laplace transform on both sides of system (4.40) yields

$$\begin{cases} s^\alpha X(s) - s^{\alpha-1} x_0 = aX(s) - bY(s), \\ s^\alpha Y(s) - s^{\alpha-1} y_0 = bX(s) + aY(s), \end{cases}$$

where $X(s) = \mathcal{L}(x(t))$, $Y(s) = \mathcal{L}(y(t))$, $x_0 = x(0)$ and $y_0 = y(0)$. It follows that

$$X(s) = \frac{(s^\alpha - a)s^{\alpha-1}x_0 - bs^{\alpha-1}y_0}{(s^\alpha - a)^2 + b^2},$$

$$Y(s) = \frac{(s^\alpha - a)s^{\alpha-1}y_0 + bs^{\alpha-1}x_0}{(s^\alpha - a)^2 + b^2},$$

that is,

$$X(s) = \frac{x_0 + y_0 i}{2}\frac{s^{\alpha-1}}{s^\alpha - a - bi} + \frac{x_0 - y_0 i}{2}\frac{s^{\alpha-1}}{s^\alpha - a + bi},$$

$$Y(s) = \frac{y_0 - x_0 i}{2}\frac{s^{\alpha-1}}{s^\alpha - a - bi} + \frac{y_0 + x_0 i}{2}\frac{s^{\alpha-1}}{s^\alpha - a + bi},$$

in which i is the imaginary unit.
Making the inverse Laplace transform and using

$$\mathcal{L}^{-1}\left(\frac{s^{\alpha-1}}{s^\alpha - a - bi}\right) = E_{\alpha,1}((a + bi)t^\alpha)$$

and

$$\mathcal{L}^{-1}\left(\frac{s^{\alpha-1}}{s^\alpha - a + bi}\right) = E_{\alpha,1}((a - bi)t^\alpha)$$

lead to

$$x(t) = \frac{x_0 + y_0 i}{2}E_{\alpha,1}((a + bi)t^\alpha) + \frac{x_0 - y_0 i}{2}E_{\alpha,1}((a - bi)t^\alpha),$$

$$y(t) = \frac{y_0 - x_0 i}{2}E_{\alpha,1}((a + bi)t^\alpha) + \frac{y_0 + x_0 i}{2}E_{\alpha,1}((a - bi)t^\alpha).$$

If $a < 0$ and $\alpha \in (0, 1)$, applying the theorem, we have

$$\lim_{t \to +\infty} x(t) = \lim_{t \to +\infty} y(t) = 0,$$

which indicates that the zero solution of system (4.40) is asymptotically stable.
In system (4.40), if one lets α approach 1 and uses $\lim_{\alpha \to 1}{}^C D^\alpha = \frac{d}{dt}$, then the original system can be changed into an ordinary differential system. For this ordinary system, the zero solution is stable for $a = 0$. However, it is not the case for fractional system (4.40).

Example 4.6.8 (*Duffing Equation*) Consider the fractional nonlinear system

$${}^C D^\alpha x(t) = -x(t) - x(t)^3, \quad 0 < \alpha < 2.$$

When $\alpha = \frac{1}{2}$, the nonlinear first-order Duffing equation is given by

$$^C D^{\frac{1}{2}} x(t) = -x(t) - x(t)^3, \tag{4.41}$$
$$x(0) = 1.$$

The homogeneous system $^C D^{\frac{1}{2}} x(t) = -x(t)$ satisfies $|arg(-1)| > \frac{\pi}{4}$. Here, $f(t, x) = -x^3$ which satisfies $f(t, 0) = 0$. Since all the conditions of Theorem 4.5.2 are satisfied, the given system (4.41) is asymptotically stable. When $\alpha = \frac{3}{2}$, the non-linear second-order Duffing equation is given by

$$^C D^{\frac{3}{2}} x(t) = -x(t) - x(t)^3, \tag{4.42}$$
$$x(0) = 1, x'(0) = 1.$$

The given equation can be converted into a system of fractional differential equations by using the substitution

$$^C D^{\frac{1}{2}} x_1(t) = x_2(t),$$
$$^C D^{\frac{1}{2}} x_2(t) = x_3(t),$$
$$^C D^{\frac{1}{2}} x_3(t) = -x_1(t) - x_1^3(t), \tag{4.43}$$

where $x_1(t) = x(t)$ with initial condition $x_1(0) = 1$, $x_2(0) = 0$, $x_3(0) = 1$. This can be written in the standard form, $^C D^{\frac{1}{2}} x(t) = Ax(t) + f(t, x)$ and $f(t, x(t)) = (0, 0, x_1^3(t))^T$ where

$$A = \begin{bmatrix} 0 & 1 & 0 \\ 0 & 0 & 1 \\ -1 & 0 & 0 \end{bmatrix}.$$

The eigenvalues of A are $-1, 0.5 \pm 0.8660i$ which satisfy $|arg(spec(A))| > \frac{\pi}{4}$. Here, the nonlinear term $f(t, x)$ satisfies $f(t, 0) = 0$. Since all the conditions of Theorem 4.5.2 are satisfied, the given system (4.42) is asymptotically stable.

Example 4.6.9 Consider the fractional nonlinear system

$$^C D^{\alpha} x_1(t) = x_2(t),$$
$$^C D^{\alpha} x_2(t) = -x_1(t) - (1 + x_2(t))^2 x_2(t), \tag{4.44}$$

with the initial condition $x_1(0) = 0.14$, $x_2(0) = 0.125$, when $\alpha = 0.9$. This can be written in the standard form $^C D^{\alpha} x(t) = Ax(t) + f(t, x)$, where $f(t, x) = (0, -(1 + x_2(t))^2 x_2(t))^T$, where

$$A = \begin{bmatrix} 0 & 1 \\ -1 & 0 \end{bmatrix}$$

The eigenvalues of A are $\pm i$ which satisfy $|arg(spec(A))| > \frac{0.8\pi}{2}$. Since the system (4.44) satisfies all the conditions of the Theorem 4.5.2, the system (4.42) is asymptotically stable.

Example 4.6.10 Consider the scalar integrodifferential equation of the form

$$^C D^\alpha x(t) + 2x(t) = -5 \int_0^t x(s)ds + h(t), \tag{4.45}$$

where $\alpha = \frac{1}{2}$ with initial condition $x(0) = 0$ and

$$h(t) = \begin{cases} 1 & 0 < t \le 1 \\ 0 & otherwise \end{cases}.$$

This can be written in the standard form as $^C D^\alpha x(t) = Ax(t) + f(t, x(t))$, where $A = -2$, $f(t, x) = \int_0^t g(t, x(s))ds + h(t)$, $g(t, x(t)) = -5x(t)$ and $h(t)$ is as above. This system (4.45) satisfies the conditions of the Theorem (4.5.5). Hence, it is asymptotically stable.

Example 4.6.11 The trivial solution of

$$^C D^\alpha x(t) = 0$$

is stable. However, if we add the perturbation $x^2(t)$ to the equation, then the solution of the perturbed equation

$$^C D^\alpha x(t) = x^2(t)$$

is unstable. On the other hand, for the perturbation $-x^3(t)$, the perturbed equation

$$^C D^\alpha x(t) = -x^3(t)$$

is asymptotically stable.

4.7 Exercises

4.1. Verify the observability of the following systems for $0 < \alpha \le 2$
  (i) $^C D^\alpha - x = 0$ with the observer $y = x$.
  (ii) $^C D^\alpha x + x = 0$ with the observer $y = \dot{x}$.

4.2. Verify the controllability of the systems $0 < \alpha \le 2$, $0 < \beta \le 1$,
  (i) $^C D^\alpha x = u$
  (ii) $^C D^\alpha x + 2b\dot{x} + x = u$

(iii) $^C D^\beta x = y + u,\ ^C D^\beta y = -x - 2by - u_1 + u_2$

(iv) $^C D^\beta x_1 = u_1 + u_2,\ ^C D^\beta x_2 = u_1 - u_2$

4.3. Prove that the system

$$\frac{d^2 x}{dt^2} = Ax + Bu, \quad x(0),\ \dot{x}(0) \in \mathbb{R}^n,$$

is controllable on $\mathbb{R}^{2n}$ iff the pair (A, B) is controllable.

4.4. Verify the controllability of the system for $0 < \beta \leq 1$

$$^C D^\beta x_1 = x_1 - x_2 + u_1 + u_2 + \frac{10x_1}{(1 + x_1^2 + u_1^2)}$$

$$^C D^\beta x_2 = x_1 + x_2 - u_1 + u_2 + \frac{x_2}{(1 + x_2^2 + u_2^2)}$$

4.5. Determine the stability of the system $^C D^\alpha x = Ax$ when

(i) $A = \begin{bmatrix} 1 & 5 \\ 5 & 1 \end{bmatrix}$, (ii) $A = \begin{bmatrix} 1 & 1 \\ -2 & -2 \end{bmatrix}$,

(iii) $A = \begin{bmatrix} -3 & -4 \\ 2 & 1 \end{bmatrix}$, (iv) $A = \begin{bmatrix} 3 & 2 & 4 \\ 2 & 0 & 2 \\ 4 & 2 & 3 \end{bmatrix}$,

(v) $A = \begin{bmatrix} 2 & -3 & 0 \\ 0 & -6 & -2 \\ -6 & 0 & -3 \end{bmatrix}$.

4.6. Analyze the stability of the fractional damped equation $^C D^\alpha x + \dot{x} + \sin x = 0$.

References

1. Balachandran, K., Dauer, J.P.: Elements of Control Theory. Narosa Publishers, New Delhi (2012)
2. Klamka, J.: Controllability of Dynamical Systems. Kluwer Academic, Dordrecht (1993)
3. Balachandran, K., Kokila, J.: On the controllability of fractional dynamical systems. Int. J. Appl. Math. Comput. Sci. 22, 523–531 (2012)
4. Balachandran, K., Kokila, J.: Constrained controllability of fractional dynamical systems. Numer. Funct. Anal. Optim. 34, 1187–1205 (2013)
5. Balachandran, K., Park, J.Y., Trujillo, J.J.: Controllability of nonlinear fractional dynamical systems. Nonlinear Anal. 75, 1919–1926 (2012)
6. Balachandran, K., Govindaraj, V., Rodriguez-Germa, L., Trujillo, J.J.: Controllability results for nonlinear fractional order dynamical systems. J. Optim. Theory Appl. 156, 33–44 (2013)
7. Balachandran, K., Govindaraj, V., Rodriguez-Germa, L., Trujillo, J.J.: Controllability of nonlinear higher order fractional dynamical systems. Nonlinear Dyn. 71, 605–612 (2013)
8. Balachandran, K., Govindaraj, V., Rodriguez-Germa, L., Trujillo, J.J.: Stabilizability of fractional dynamical systems. Fract. Calc. Appl. Anal. 17, 511–531 (2014)

9. Balachandran, K., Govindaraj, V., Rivero, M., Trujillo, J.J.: Controllability of fractional damped dynamical systems. Appl. Math. Comput. **257**, 66–73 (2015)
10. Chen, L., He, Y., Chai, Y., Wu, R.: New results on stability and stabilization of a class of nonlinear fractional-order systems. Nonlinear Dyn. **75**, 633–641 (2014)
11. Kaczorek, T.: Selected Problems of Fractional Systems Theory. Springer, Berlin (2011)
12. Li, Y., Chen, Y., Podlubny, I.: Mittag-Leffler stability of fractional order nonlinear dynamic systems. Automatica **45**, 1965–1969 (2009)
13. Monje, C.A., Chen, Y.Q., Vinagre, B.M., Xue, X., Feliu, V.: Fractional-Order Systems and Controls: Fundamentals and Applications. Springer, London (2010)
14. Matignon, D., d'Andrea-Novel, B.: Some results on controllability and observability of finite-dimensional fractional differential systems. Comput. Eng. Syst. Appl. **2**, 952–956 (1996)
15. Podlubny, I.: Fractional Differential Equations. Academic Press, San Diego (1999)
16. Matignon, D.: Stability results for fractional differential equations with applications to control processing. Comput. Eng. Syst. Appl. **2**, 963–968 (1996)
17. Barbu, V.: Differential Equations. Springer, Switzerland (2016)
18. Agarwal, R., Wong, J.Y., Li, C.: Stability analysis of fractional differential system with Riemann-Liouville derivative. Math. Comput. Model. **52**, 862–874 (2010)

Chapter 5
Fractional Partial Differential Equations

Abstract Fractional partial differential equations encountered in different fields of study are given and the concepts of fractional partial integral and fractional partial derivative are introduced. Explicit solutions for standard linear fractional partial differential equations are obtained. Existence of solutions of nonlinear fractional partial differential equations with different boundary conditions is established with the help of Leray–Schauder's fixed point theorem. Examples are constructed to illustrate the theory and a set of exercises is added.

Keywords Fractional partial integral · Fractional partial derivative · Exact solutions · Existence of solutions · Fractional partial differential equations · Leray–Schauder's theorem

5.1 Motivation

Fractional partial differential equations are found to be an effective tool to describe certain physical phenomena such as diffusion processes and viscoelasticity. In particular, engineers and scientists have developed innovative models that involve fractional partial differential equations and have applied successfully in the modeling of polymers and proteins, transmission of ultrasound waves, modeling of human tissue under mechanical loads, and many other areas of science. In recent years, theoretical analysis and the numerical methods of fractional partial differential equations have attracted many researchers. Due to increasing interest, a lot of researchers are studying fractional partial differential equations. Also fractional partial differential equations are more and more helpful in modeling fluid flow and anomalous diffusion in biological systems and nonlinear fractional partial differential equations describe many physical phenomena in a better way. In this chapter, we present few applications of fractional derivatives in modeling biological and physical phenomena and solve few basic fractional equations. Further we prove certain existence theorems for nonlinear fractional partial differential equations by means of fixed point theorem.

Fractional partial differential equations have been encountered in different fields of study. Here we provide few models occurring in some real-world problems.

© The Author(s), under exclusive license to Springer Nature Singapore Pte Ltd. 2023   115
K. Balachandran, *An Introduction to Fractional Differential Equations*, Industrial and Applied Mathematics, https://doi.org/10.1007/978-981-99-6080-4_5

(i) Fractional Diffusion Model

Fractional diffusion equations were first introduced by Nigmatullin [1] in fractal media to analyze the thermal diffusion. Some random motion in financial systems confirms a model with subdiffusion such as fractional diffusion model. In a subdiffusion process, the growth of mean square variance is slower than that of a Gaussian process. This dynamic feature is modeled in terms of fractional diffusion equation as

$$\frac{{}^{C}\partial^{\alpha}u(x,t)}{\partial t^{\alpha}} - \nabla\big(a(x,t)\nabla u(x,t)\big) = f(x,t), \quad \text{on } \Omega \times (0,T].$$

Here Ω is a convex bounded polygonal domain in $\mathbb{R}^n$, $n = 1, 2, 3$, and $T > 0$. These fractional diffusion models are also used in the dynamics of chemical reactions and phase transitions that have subdiffusion process.

(ii) Fractional Fisher–Kolmogorov Equation

The Fisher–Kolmogorov equation is one of the nonlinear reaction diffusion equations, which was formulated by Fisher in 1937 and its analytical results were developed by Kolmogorov. At the beginning, it was introduced to simulate the propagation of a gene within a population and the solution of Fisher–Kolmogorov equation is a wave that travels in space and time with an obverse that never changes its shape. The time fractional Fisher–Kolmogorov equation is attained by changing the time derivative of the Fisher–Kolmogorov equation with a fractional derivative of order α, $0 < \alpha < 1$. The extended time fractional Fisher–Kolmogorov equation is of the form

$$\frac{{}^{C}\partial^{\alpha}u}{\partial t^{\alpha}} - \Delta u + \Delta^2 u + u^3 - u = 0, \quad 0 < \alpha < 1,$$

where $u(x,t)$ denotes the density of particles or individuals. On discussion regarding epidemics, $u(x,t)$ could mean the density of infected individuals (see [2]).

(iii) Fractional Cable Equation

Diffusion is a most important transport mechanism in the living organism. Since the diffusion is anomalous in disordered media, to model those anomalous subdiffusion the integer time derivative is replaced by fractional time derivative in the corresponding models such as

$$\frac{{}^{C}\partial^{\alpha}u}{\partial t^{\alpha}} - \frac{\partial^2 u}{\partial x^2} + u(x,t) = 0.$$

For example, through MRI, the anomalous diffusion is frequently detected in brain tissues. In order to find a model of the brain, we should investigate how the voltage propagates in brain tissues. To be precise, it is important to understand how the voltage propagates in a cable with anomalous diffusion because axons in neuron can be expressed as cables. Those are very important in the neuron–neuron communications. In this regard, more recently authors have suggested the fractional cable equation [3] of the form

$$M\frac{\partial v}{\partial t} = \beta \frac{^C\partial^{1-\alpha}}{\partial t^{1-\alpha}}\left(\frac{d}{4r_L}\frac{\partial^2 v}{\partial x^2} - i_c\right), \quad 0 \leq \alpha \leq 1,$$

where $v(x,t)$ is the voltage in a cylindrical cable, M represents the membrane capacitance, d is a diameter of a cylindrical cable, r_L indicates the longitudinal resistance, β is a constant, and i_c specifies the ionic current flowing per unit area in the cable both inward and outward directions. Here the Riemann–Liouville fractional derivative is considered. In nerve cells to analyze the electrodiffusion of ions, the current i_c is taken as v/r_M, where r_M is the membrane resistance. Since fractals are used to derive stretched exponential function in a brain tissue, the diffusion is described by this function in [4].

(iv) Fractional Burgers Equation

Burgers equation is the simplest nonlinear equation for diffusive waves in fluid dynamics. We investigate Burgers equation with a fractional time derivative

$$\frac{\partial u}{\partial t} + \frac{\partial}{\partial x}\left(au + b\frac{u^2}{2}\right) = -\epsilon\frac{^C\partial^\alpha u}{\partial t^\alpha}, \quad \epsilon \geq 0, \ 0 < \alpha < 1, \tag{5.1}$$

where a is a constant speed of advection and b is a coefficient of nonlinear term. When $\epsilon = 0$, the above equation reduces to a transport equation. The fractional time derivative term on the right-hand side describes the effect of memory and linear losses during the propagation of waves. Similarly, when the model has anomalous diffusion or dispersion or sedimentation of particles, the space fractional Burger's equation has been referred by the authors [5] which is obtained by replacing fractional Laplacian rather than fractional time derivative in the right-hand side of (5.1). In a gas-filled pipe, the physical processes of acoustic waves which have unidirectional propagation are modeled by using fractional derivatives as

$$\frac{^C\partial^\alpha u}{\partial t^\alpha} + \epsilon u\frac{\partial u}{\partial x} - \mu\frac{\partial^2 u}{\partial x^2} + \eta\frac{^C\partial^\beta u}{\partial x^\beta} = 0, \quad 0 < \alpha, \beta \leq 1, \ t > 0,$$

where ϵ, μ, η are the parameters. Such equations also appear in shallow-water waves and waves in bubbly liquids [5, 6].

(v) Fractional Population Model

The fractional population model is of the form

$$\frac{{}^C\partial^\alpha u}{\partial t^\alpha} = \frac{\partial^2 u}{\partial x^2} + \frac{\partial^2 u}{\partial y^2} + f(u).$$

Here $u(x, t)$ represents the population density and f denotes the population supply due to births and deaths. This equation is used to calculate the dynamics of the population changes. Gurney and Nisbet [7] considered the animal population model, wherein the population density of the animal species moves from higher population to lower population density. They observed that the movement or migration of the species takes place at a faster rate at higher density populations when compared to the lower densities. To model this scenario, they considered a fine rectangular mesh in such a way that the animal species are located at the grid points in the mesh. The population model was then formulated by incorporating the nature of migration as for each time step, the animal either remains at its location or moves toward lower population density locations. The probability of such movement of the species depends on the magnitude of the density gradient in the considered mesh. This approach leads to the following equation:

$$\frac{{}^C\partial^\alpha u}{\partial t^\alpha} = \frac{\partial^2 u^2}{\partial x^2} + \frac{\partial^2 u^2}{\partial y^2} + f(u).$$

For more details about the fractional partial differential equations, the reader can refer the books [8, 9] and the papers [10, 11].

5.2   Fractional Partial Integral and Derivative

We introduce the following definitions from fractional calculus to study the fractional partial differential equations.

Definition 5.2.1 [12] The Riemann–Liouville fractional partial integral operator of order α with respect to t of a function $f(x, t)$ is defined by

$$I^\alpha f(x, t) = \frac{1}{\Gamma(\alpha)} \int_0^t \frac{f(x, s)}{(t - s)^{1-\alpha}} \, ds,$$

where $f(\cdot, t)$ is an integrable function.

For any $n - 1 < \alpha < n$, $n \in \mathbb{N}$, the Riemann–Liouville and Caputo fractional partial derivative operators are defined as follows:

Definition 5.2.2 [12] The Riemann–Liouville fractional partial derivative of order α of a function $f(x, t)$ with respect to t is defined by

$$\frac{\partial^{\alpha} f(x, t)}{\partial t^{\alpha}} = \frac{1}{\Gamma(n - \alpha)} \frac{\partial^{n}}{\partial t^{n}} \int_{0}^{t} \frac{f(x, s)}{(t - s)^{\alpha - n + 1}} \, ds,$$

where the function $f(\cdot, t)$ has absolutely continuous derivatives up to order $(n - 1)$.

Definition 5.2.3 [12] The Caputo fractional partial derivative of order α with respect to t of a function $f(x, t)$ is defined as

$$\frac{{}^{C}\partial^{\alpha} f(x, t)}{\partial t^{\alpha}} = \frac{1}{\Gamma(n - \alpha)} \int_{0}^{t} \frac{1}{(t - s)^{\alpha - n + 1}} \frac{\partial^{n} f(x, s)}{\partial s^{n}} \, ds,$$

where the function $f(\cdot, t)$ has absolutely continuous derivatives up to order $(n - 1)$.

The Riemann–Liouville and Caputo fractional partial derivatives are linked by the following relationship:

$$\frac{{}^{C}\partial^{\alpha} f(x, t)}{\partial t^{\alpha}} = \frac{\partial^{\alpha} f(x, t)}{\partial t^{\alpha}} - \sum_{k=0}^{n-1} \frac{t^{k - \alpha}}{\Gamma(k + 1 - \alpha)} \frac{\partial^{k} f(x, 0)}{\partial t^{k}}.$$

Although the function with Riemann–Liouville fractional partial derivative need not be continuous at the origin and differentiable, it has some disadvantages when dealing with real-world problem. Specifically, while using Riemann–Liouville derivative, the fractional partial differential equation needs the fractional order initial condition and the Riemann–Liouville fractional partial derivative of a constant K is $\frac{\partial^{\alpha} K}{\partial t^{\alpha}} = Kt^{-\alpha}/\Gamma(1 - \alpha)$. But fractional partial derivative in Caputo sense is applicable when trying to model real-world phenomena because Caputo partial derivative admits conventional initial conditions and the Caputo fractional partial derivative of a constant is zero as in integer order case. Both the fractional partial derivatives coincide with zero initial condition.

5.3 Linear Fractional Equations

Consider the initial value problem

$$\frac{{}^{C}\partial^{\alpha} u(x, t)}{\partial t^{\alpha}} = f(x, t), \quad t \in J = [0, T], \tag{5.2}$$

$$u(x, 0) = u_0(x)$$

where $0 < \alpha < 1$ and $f : \mathbb{R} \times J \to \mathbb{R}$ is a continuous function. Then the corresponding integral equation is

$$u(x, t) = u_0(x) + \frac{1}{\Gamma(\alpha)} \int_0^t \frac{f(x, s)}{(t - s)^{1-\alpha}} \, ds. \tag{5.3}$$

We observe the following two cases:

1. If $f(\cdot, t) \in L^1([0, T])$ or $f(\cdot, t) \in C([0, T])$, then the derivative of $u(x, t)$ does not exist. Hence $\frac{{}^C \partial^\alpha u(x,t)}{\partial t^\alpha}$ does not exist and the equivalent relation between (5.2) and (5.3) is not possible. In this case, (5.3) is not a strong solution but called a mild solution of (5.2).
2. If $f(\cdot, t) \in AC([0, T])$ (the space of all absolutely continuous functions), then $u(\cdot, t) \in AC([0, T])$. This shows the equivalent relation between (5.2) and (5.3) and the problem (5.2) can be solved for a strong solution.

Lemma 5.3.1 (Green's Identity) *[13] Let Ω be a bounded domain in $\mathbb{R}^m$ with smooth boundary $\partial\Omega$. Then, for any $u, v \in C^2(\Omega)$,*

$$\int_\Omega v \Delta u \, dx = \int_{\partial\Omega} v \frac{\partial u}{\partial n} \, ds - \int_\Omega \nabla u \cdot \nabla v \, dx,$$

where n is the outward unit normal to the boundary $\partial\Omega$ and ds is the element of arc length. For the special case $v = 1$,

$$\int_\Omega \Delta u \, dx = \int_{\partial\Omega} \frac{\partial u}{\partial n} \, ds. \tag{5.4}$$

This is called Green's first identity.

5.3.1   Adomian Decomposition Method

To describe this method [14], consider the differential equation of the form

$$Lu + Ru + Nu = f, \tag{5.5}$$

where L is a linear operator which can be inverted, R is the remainder operator, N is a nonlinear operator, and f is the source term. Since L is invertible, we have from (5.5),

$$u = L^{-1}Lu = L^{-1}f - L^{-1}Ru - L^{-1}Nu. \tag{5.6}$$

The beauty of the method is that the solution is first expressed as a series and the terms are approximated one by one. To be clear, suppose u can be expressed as $u = \sum\limits_{n=0}^{\infty} u_n$ with $u_0 = \phi + L^{-1} f$, where ϕ denotes terms appearing from the initial or boundary conditions. The nonlinear term Nu can be written in terms of Adomian polynomials as

$$Nu = \sum_{n=0}^{\infty} A_n,$$

where A_n is calculated using the formula

$$A_n = \frac{1}{n!} \left[\frac{d^n}{d\lambda^n} N(v(y)) \right]_{\lambda=0}, \quad n = 0, 1, 2, \ldots,$$

and

$$v(y) = \sum_{n=0}^{\infty} \lambda^n u_n.$$

Now the series solution $u = \sum\limits_{n=0}^{\infty} u_n$ to the differential equation is calculated iteratively as follows:

$$u_0 = \phi + L^{-1} f$$
$$u_{n+1} = -L^{-1} R u_n - L^{-1} A_n, \quad n \geq 0,$$

and then summing up all the terms to obtain u. This method can be extended to partial differential equations and fractional differential equations. Suppose we consider the partial differential equation

$$L_t u = L_x u + f,$$

with initial condition $u(x, 0) = g(x)$, where L_t and L_x are differential operators with respect to t and x, respectively. Then the form of (5.6) in this case would be

$$L_t^{-1} L_t u = L_t^{-1} (L_x u + f).$$

Therefore the solution algorithm is

$$u_0(x, t) = g(x) + L_t^{-1} f,$$
$$u_{n+1}(x, t) = L_t^{-1} L_x u_n, \quad n \geq 0.$$

For the heat equation $L_t^{-1} = \int_0^t (\cdot) \, ds$ and for the wave equation $L_t^{-1} = \int_0^{t_1} \int_0^{t_2} (\cdot) \, ds_2 \, ds_1$. In the case of fractional differential equation, L_t^{-1} is the Riemann–Liouville fractional integral operator and is denoted by I^α. Application of this method to different fractional partial differential can be found in [10].

5.3.2 Fractional Diffusion Equation

Consider the fractional diffusion equation of the form

$$\frac{{}^C \partial^\alpha u}{\partial t^\alpha} = \kappa \frac{{}^C \partial^\beta u}{\partial x^\beta}, \quad \begin{array}{c} 0 < \alpha \leq 1, \ 1 < \beta \leq 2, \\ 0 < x < l, \ t > 0, \end{array} \tag{5.7}$$

with the initial and boundary conditions

$$u(0, t) = 0 = u(l, t), t \geq 0, \tag{5.8}$$

$$u(x, 0) = f(x), 0 \leq x \leq l, \tag{5.9}$$

where κ is a constant (diffusivity constant in heat diffusion problems).

We assume a nonzero separable solution of (5.7) in the form

$$u(x, t) = X(x)T(t). \tag{5.10}$$

Substituting (5.10) in (5.7) gives (here ${}^C D^\eta = \frac{{}^C d^\eta}{dy^\eta}$ where $\eta = \alpha$ or β and $y = x$ or t)

$$\frac{1}{X} \frac{{}^C d^\beta X}{dx^\beta} = \frac{1}{\kappa T} \frac{{}^C d^\alpha T}{dt^\alpha}. \tag{5.11}$$

Since the left-hand side depends only on x and the right-hand side is a function of time t only, result (5.11) can be true only if both sides are equal to the same constant λ. Thus we obtain two ordinary fractional differential equations

$$\frac{{}^C d^\beta X}{dx^\beta} - \lambda X = 0 \text{ and } \frac{{}^C d^\alpha T}{dt^\alpha} - \lambda \kappa T = 0. \tag{5.12}$$

We consider λ as negative, since all other values are not of physical interest. Hence let $\lambda = -\gamma^2$. Therefore Eq. (5.12) becomes

$$\frac{{}^C d^\beta X}{dx^\beta} + \gamma^2 X = 0, \text{ and } \frac{{}^C d^\alpha T}{dt^\alpha} + \kappa \gamma^2 T = 0. \tag{5.13}$$

The above equations can be solved by means of Laplace transform technique [12] and the solution in terms of Mittag-Leffler function is given as

$$X(x) = A E_{\beta,1}(-\gamma^2 x^\beta) + Bx E_{\beta,2}(-\gamma^2 x^\beta) \tag{5.14}$$

and

$$T(t) = C E_\alpha(-\kappa \gamma^2 t^\alpha), \tag{5.15}$$

where A, B, and C are arbitrary constants.

The boundary conditions for $X(x)$ are

$$X(0) = 0 = X(l) \tag{5.16}$$

which are used to find A and B in the solution (5.14). It turns out that $A = 0$ and $B \neq 0$. Hence

$$Bl E_{\beta,2}(-\gamma^2 l^\beta) = 0 \tag{5.17}$$

gives the eigenvalues and the corresponding eigenfunctions are given by

$$X_n(x) = B_n x E_{\beta,2}(-\gamma_n^2 x^\beta), \tag{5.18}$$

where B_n are constants.

With $\gamma = \gamma_n$, we obtain the solution $u_n(x, t)$ as

$$u_n(x, t) = a_n E_\alpha(-\gamma_n^2 \kappa t) x E_{\beta,2}(-\gamma_n^2 x^\beta), \tag{5.19}$$

where $a_n = B_n C_n$. Thus the most general solution is obtained by the principle of superposition in the form

$$u(x, t) = \sum_{n=1}^{\infty} a_n E_\alpha(-\gamma_n \kappa t^\alpha) x E_{\beta,2}(-\gamma_n x^\beta). \tag{5.20}$$

The constant a_n is calculated by using the initial condition $u(x, 0) = f(x)$.

5.3.3 Fractional Wave Equation

Consider the one-dimensional fractional wave equation of the form

$$\frac{{}^C\partial^\alpha u(x, t)}{\partial t^\alpha} = c^2 \frac{{}^C\partial^\beta u(x, t)}{\partial x^\beta}, \quad \begin{matrix} 0 \leq x \leq l, \ t \geq 0, \\ 1 < \alpha \leq 2, \ 1 < \beta \leq 2, \end{matrix} \tag{5.21}$$

with initial and boundary conditions

$$u(0, t) = 0 = u(l, t), t \geq 0,$$
$$u(x, 0) = f(x), u_t(x, 0) = g(x) \ 0 \leq x \leq l, \tag{5.22}$$

where c^2 is vibrating string constant.

We assume a separable solution of (5.21) in the form

$$u(x, t) = X(x)T(t). \tag{5.23}$$

Substitution of (5.23) in (5.21) gives

$$\frac{1}{c^2 T} \frac{^C d^\alpha T}{dt^\alpha} = \frac{1}{X} \frac{^C d^\beta X}{dx^\beta}. \tag{5.24}$$

Since the left-hand side depends only on t and the right-hand side is a function of time x only, (5.24) can be true only if both sides are equal to the same constant μ. Thus we obtain two ordinary fractional differential equations written as

$$\frac{1}{c^2 T} \frac{^C d^\alpha T}{dt^\alpha} = \frac{1}{X} \frac{^C d^\beta X}{dx^\beta} = \mu. \tag{5.25}$$

Thus the determination of solutions of the original fractional partial differential equation has been reduced to the determination of solutions of the two fractional differential equations

$$\frac{^C d^\alpha T}{dt^\alpha} = c^2 \mu T \quad \text{and} \quad \frac{^C d^\beta X}{dx^\alpha} = \mu X. \tag{5.26}$$

The cases $\mu > 0$ and $\mu = 0$ lead to trivial zero solution and hence we take $\mu < 0$; we write $\mu = -\lambda$. Then the component differential equations and their solutions are

$$X(x) = A E_{\beta,1}(-\lambda x^\beta) + B x E_{\beta,2}(-\lambda x^\beta). \tag{5.27}$$

From the condition $X(x) = 0$, we obtain $A = 0$. The condition $X(l) = 0$ gives

$$Bl E_{\beta,2}(-\lambda l^\beta) = 0.$$

Solving the above equation, we get the eigenvalues λ_n and the corresponding eigenfunctions are given by

$$X_n(x) = B_n x E_{\beta,2}(-\lambda_n x^\beta).$$

Now for $\lambda = \lambda_n$, the general solution of equation

$$\frac{^C d^\alpha T}{dt^\alpha} = c^2 \mu T$$

is given by

$$T_n(t) = C_n E_{\alpha,1}(-c^2 \lambda_n t^\alpha) + D_n t E_{\alpha,2}(-c^2 \lambda_n t^\alpha).$$

Hence the general solution is given by

$$u_n(x, t) = \left(a_n E_{\alpha,1}(-c^2 \lambda_n t^\alpha) + b_n t E_{\alpha,2}(-c^2 \lambda_n t^\alpha)\right) x E_{\beta,2}(-\lambda_n x^\beta),$$

where $a_n = B_n C_n$ and $b_n = B_n D_n$. Since the equation is linear and homogeneous, by the principle of superposition

$$u(x, t) = \sum_{n=1}^{\infty} \left(a_n E_{\alpha,1}(-c^2 \lambda_n t^\alpha) + b_n t E_{\alpha,2}(-c^2 \lambda_n t^\alpha)\right) x E_{\beta,2}(-\lambda_n x^\beta)$$

is also a solution. By using the remaining two conditions on $f(x)$ and $g(x)$, we calculate the constants a_n and b_n. The eigenvalues for different values of β can be easily obtained and if we take $\alpha = \beta = 2$, the above solution reduces to

$$u(x, t) = \sum_{n=1}^{\infty} \left(a_n \cos(\frac{n\pi c}{l}t) + b_n \sin(\frac{n\pi c}{l}t)\right) \sin(\frac{n\pi x}{l}),$$

which is the solution to the classical wave equation.

5.3.4 Fractional Black–Scholes Equation

Consider the fractional Black–Scholes option pricing equation [15, 16] of the form

$$\frac{{}^C \partial^\alpha v}{\partial t^\alpha} = \frac{\partial^2 v}{\partial x^2} + (k - 1)\frac{\partial v}{\partial x} - kv, \quad 0 < \alpha \le 1, \tag{5.28}$$

with the initial condition $v(x, 0) = e^x - 1$, $x > 0$.

For applying the Adomian decomposition method, we rewrite Eq. (5.28) as

$$\frac{\partial v}{\partial t} = \frac{{}^C \partial^{1-\alpha}}{\partial t^{1-\alpha}} \left[\frac{\partial^2 v}{\partial x^2} + (k - 1)\frac{\partial v}{\partial x} - kv\right].$$

Now integrating on both sides with respect to t, we get

$$v(x, t) = v(x, 0) + \int_0^t \left(\frac{{}^C \partial^{1-\alpha}}{\partial s^{1-\alpha}} \left[\frac{\partial^2 v(x, s)}{\partial x^2} + (k - 1)\frac{\partial v(x, s)}{\partial x} - kv(x, s)\right]\right) ds.$$

Here we take $v(x, 0)$ as v_0 and

$$v_1(x, t) = \int_0^t \left(\frac{{}^C\partial^{1-\alpha}}{\partial s^{1-\alpha}} \left[\frac{\partial^2 v_0(x, s)}{\partial x^2} + (k - 1)\frac{\partial v_0(x, s)}{\partial x} - kv_0(x, s) \right] \right) ds.$$

The iterative scheme, for $n > 1$, to this problem is given by

$$v_n(x, t) = \int_0^t \left(\frac{{}^C\partial^{1-\alpha}}{\partial s^{1-\alpha}} \left[\frac{\partial^2 v_{n-1}(x, s)}{\partial x^2} + (k - 1)\frac{\partial v_{n-1}(x, s)}{\partial x} - kv_{n-1}(x, s) \right] \right) ds.$$

(5.29)

Hence, by Adomian decomposition method, the solution to (5.28) is given by

$$v(x, t) = \sum_{n=0}^{\infty} v_n(x, t).$$

Here we consider the initial condition $v(x, 0) = e^x - 1, \ x > 0$; hence $v_0 = e^x - 1$. By using (5.29), we calculate the remaining terms of the solution as

$$v_1(x, t) = \int_0^t \left(\frac{{}^C\partial^{1-\alpha}}{\partial s^{1-\alpha}} \left[\frac{\partial^2 v_0(x, 0)}{\partial x^2} + (k - 1)\frac{\partial v_0(x, 0)}{\partial x} - kv_0(x, 0) \right] \right) ds$$

$$= -e^x \frac{-kt^\alpha}{\Gamma(\alpha + 1)} + (e^x - 1)\frac{-kt^\alpha}{\Gamma(\alpha + 1)}$$

$$v_2(x, t) = \int_0^t \left(\frac{{}^C\partial^{1-\alpha}}{\partial s^{1-\alpha}} \left[\frac{\partial^2 v_1(x, 0)}{\partial x^2} + (k - 1)\frac{\partial v_1(x, 0)}{\partial x} - kv_1(x, 0) \right] \right) ds$$

$$= -e^x \frac{(-kt^\alpha)^2}{\Gamma(2\alpha + 1)} + (e^x - 1)\frac{(-kt^\alpha)^2}{\Gamma(2\alpha + 1)}$$

and so on. So, by Adomian decomposition method, the solution of Eq. (5.28) is given by

$$v(x, t) = \sum_{n=0}^{\infty} v_n(x, t)$$

$$= e^x \left(\frac{kt^\alpha}{\Gamma(\alpha + 1)} - \frac{(kt^\alpha)^2}{\Gamma(2\alpha + 1)} + \cdots \right)$$

$$+ (e^x - 1)\left(1 - \frac{kt^\alpha}{\Gamma(\alpha + 1)} + \frac{(kt^\alpha)^2}{\Gamma(2\alpha + 1)} + \cdots \right)$$

$$= e^x (1 - E_\alpha(-kt^\alpha)) + (e^x - 1)E_\alpha(-kt^\alpha),$$

$$= e^x - E_\alpha(-kt^\alpha),$$

(5.30)

where $E_\alpha(x)$ is Mittag-Leffler function in one parameter. Equation (5.30) is the exact solution of Eq. (5.28).

5.4 Nonlinear Fractional Equations

Consider the nonlinear fractional partial differential equation of the form

$$\frac{{}^c\partial^\alpha u}{\partial t^\alpha} = a(t)\Delta u(x,t) + f(t, u(x,t)), \quad t \in J, \tag{5.31}$$

with initial condition

$$u(x,0) = u_0(x), \qquad x \in \Omega,$$

where $0 < \alpha < 1$, Ω is a bounded subset of $\mathbb{R}$ with smooth boundary $\partial\Omega$, $J = [0, T]$, and $f : J \times \mathbb{R} \to \mathbb{R}$ is a nonlinear continuous function.
We assume the following hypotheses to prove our main result:

(H1) $a(t)$ is continuous on J and there exists a $\beta \in (0, \alpha)$ such that $a(t) \in L^{1/\beta}(0, T)$
and $\left(\int_0^T (a(s))^{\frac{1}{\beta}} \, ds \right)^\beta \leq C_1$ for $C_1 > 0$.

(H2) $f(t, u)$ is continuous with respect to u, Lebesgue measurable with respect to t and satisfies

$$\frac{\int_\Omega \phi(x) f(t, u) \, dx}{\int_\Omega \phi(x) \, dx} \leq f\left(t, \frac{\int_\Omega \phi(x) u(x,t) \, dx}{\int_\Omega \phi(x) \, dx} \right),$$

for some function $\phi(x)$.

(H3) There exists an integrable function $m(t) : J \to [0, \infty)$ such that

$$\| f(t, u) \| \leq m(t) \|u\|,$$

where $m(t) \in L^{1/\beta}(0, T)$, for some $\beta \in (0, \alpha)$, and

$$\left(\int_0^T (m(s))^{\frac{1}{\beta}} \, ds \right)^\beta \leq C_2, \text{ for } C_2 > 0.$$

It is easy to show that the initial value problem (5.31) is equivalent to the following integral equation:

$$u(x,t) = u_0(x) + \frac{1}{\Gamma(\alpha)} \int_0^t (t-s)^{\alpha-1} a(s) \Delta u(x,s)\, ds$$

$$+ \frac{1}{\Gamma(\alpha)} \int_0^t (t-s)^{\alpha-1} f(s, u(x,s))\, ds, \tag{5.32}$$

for $t \in J$.

5.4.1 Dirichlet Boundary Condition

Now we prove the existence of solution of (5.31) with Dirichlet boundary condition

$$u(x,t) = 0, \qquad (x,t) \in \partial\Omega \times J, \tag{5.33}$$

where $\partial\Omega$ is the boundary of Ω. In order to prove the result, consider the following eigenvalue problem:

$$\left. \begin{array}{l} \Delta u + \lambda u = 0, \ (x,t) \in \Omega \times J, \\ u = 0, \ (x,t) \in \partial\Omega \times J, \end{array} \right\} \tag{5.34}$$

where λ is a constant not depending on the variables x and t. Thus for $x \in \Omega$, the smallest eigenvalue λ_1 of the problem (5.34) is positive and the corresponding eigenfunction $\phi(x) \geq 0$. Define the function $U(t)$ as

$$U(t) = \frac{\int_\Omega u(x,t)\phi(x)\, dx}{\int_\Omega \phi(x)\, dx} \tag{5.35}$$

and for any constant $b > 0$, take

$$r_1 = \min\left\{ T, \left[\frac{\Gamma(\alpha)b}{(\|U(0)\| + b)(\lambda_1 C_1 + C_2)} \left(\frac{\alpha - \beta}{1 - \beta} \right)^{1-\beta} \right]^{\frac{1}{\alpha-\beta}} \right\}.$$

Theorem 5.4.1 *Assume that there exists $\beta \in (0, \alpha)$ for some $\alpha > 0$ such that (H1)–(H3) hold. Then there exists at least one solution for the initial value problem (5.31) on $\Omega \times [0, r_1]$.*

Proof First we have to prove that the initial value problem (5.31) with condition (5.33) has a solution if and only if the equation

$$U(t) = U(0) - \frac{\lambda_1}{\Gamma(\alpha)} \int_0^t (t-s)^{\alpha-1} U(s)\, ds + \frac{1}{\Gamma(\alpha)} \int_0^t (t-s)^{\alpha-1} f(s, U(s))\, ds$$

has a solution.

Step 1. Suppose $U(t)$ is a solution of the above equation. Then by Lemma 3.1 of [17] $u(x, t)$ is a solution of Eq. (5.32). Conversely, let $u(x, t)$ be a solution of (5.31). This implies that $u(x, t)$ is a solution of (5.32). Now multiplying both sides of Eq. (5.32) by $\phi(x)$ and integrating with respect to $x \in \Omega$, we get

$$\int_\Omega \phi(x) u(x, t) \, dx = \int_\Omega \phi(x) u_0(x) \, dx + \frac{1}{\Gamma(\alpha)} \int_\Omega \phi(x) \int_0^t (t - s)^{\alpha-1} a(s) \Delta u(x, s) \, ds \, dx$$
$$+ \frac{1}{\Gamma(\alpha)} \int_\Omega \phi(x) \int_0^t (t - s)^{\alpha-1} f(s, u(x, s)) \, ds \, dx.$$

Invoking Assumption (H2), we get

$$U(t) \leq U(0) - \frac{\lambda_1}{\Gamma(\alpha)} \int_0^t (t - s)^{\alpha-1} a(s) U(s) \, ds + \frac{1}{\Gamma(\alpha)} \int_0^t (t - s)^{\alpha-1} f(s, U(s)) \, ds.$$

Let $K = \{U : U \in C(J, \mathbb{R}), \| U(t) - U(0) \| \leq b\}$ for some $b > 0$. Define an operator $F : C(J, \mathbb{R}) \to C(J, \mathbb{R})$ by

$$FU(t) = U(0) - \frac{\lambda_1}{\Gamma(\alpha)} \int_0^t (t - s)^{\alpha-1} a(s) U(s) \, ds + \frac{1}{\Gamma(\alpha)} \int_0^t (t - s)^{\alpha-1} f(s, U(s)) \, ds.$$

Clearly $U(0) \in K$. This means that K is nonempty. From our construction of K, we can say that K is closed and bounded. Now, for any $U_1, U_2 \in K$ and for any $a_1, a_2 \geq 0$ such that $a_1 + a_2 = 1$,

$$\| a_1 U_1(t) + a_2 U_2(t) - U(0) \| \leq a_1 \| U_1(t) - U(0) \| + a_2 \| U_2(t) - U(0) \|$$
$$\leq a_1 b + a_2 b = b.$$

Thus $a_1 U_1 + a_2 U_2 \in K$. Therefore K is a nonempty closed convex set. Next we have to prove that the operator F maps K into itself.

$$\| FU(t) - U(0) \| = \left\| \frac{\lambda_1}{\Gamma(\alpha)} \int_0^t (t - s)^{\alpha-1} a(s) U(s) \, ds \right.$$
$$\left. + \frac{1}{\Gamma(\alpha)} \int_0^t (t - s)^{\alpha-1} f(s, U(s)) \, ds \right\|$$
$$\leq \frac{\lambda_1}{\Gamma(\alpha)} (\|U(0)\| + b) \int_0^t (t - s)^{\alpha-1} \|a(s)\| \, ds + \frac{1}{\Gamma(\alpha)} \int_0^t (t - s)^{\alpha-1} \|f(s, U(s))\| \, ds.$$

Then, by using Assumptions (H1), (H3) and Holder's inequality, we achieve

$$\| FU(t) - U(0) \|$$

$$\leq \frac{\lambda_1}{\Gamma(\alpha)} (\|U(0)\| + b) \int_0^t (t - s)^{\alpha-1} \|a(s)\| ds + \frac{1}{\Gamma(\alpha)} \int_0^t m(s)(t - s)^{\alpha-1} \|U(s)\| ds$$

$$\leq \frac{\lambda_1}{\Gamma(\alpha)} (\|U(0)\| + b) \left(\int_0^t \left((t - s)^{\alpha-1} \right)^{\frac{1}{1-\beta}} ds \right)^{1-\beta} \left(\int_0^t \|a(s)\|^{\frac{1}{\beta}} ds \right)^{\beta}$$

$$+ \frac{1}{\Gamma(\alpha)} (\|U(0)\| + b) \left(\int_0^t \left((t - s)^{\alpha-1} \right)^{\frac{1}{1-\beta}} ds \right)^{1-\beta} \left(\int_0^t (m(s))^{\frac{1}{\beta}} ds \right)^{\beta}$$

$$\leq \frac{(\|U(0)\| + b) \lambda_1 C_1}{\Gamma(\alpha)} \left(\frac{1 - \beta}{\alpha - \beta} \right)^{1-\beta} r_1^{\alpha-\beta} + \frac{(\|U(0)\| + b) C_2}{\Gamma(\alpha)} \left(\frac{1 - \beta}{\alpha - \beta} \right)^{1-\beta} r_1^{\alpha-\beta}$$

$$= \frac{(\|U(0)\| + b) (\lambda_1 C_1 + C_2)}{\Gamma(\alpha)} \left(\frac{1 - \beta}{\alpha - \beta} \right)^{1-\beta} r_1^{\alpha-\beta} \leq b.$$

Therefore F maps K into itself. Now define a sequence $\{U_k(t)\}$ in K such that

$$U_0(t) = U(0) \quad \text{and} \quad U_{k+1}(t) = FU_k(t), \quad k = 0, 1, 2, \ldots$$

Since K is closed, there exists a subsequence $\{U_{k_i}(t)\}$ of $U_k(t)$ and $\widetilde{U}(t) \in K$ such that

$$\lim_{k_i \to \infty} U_{k_i}(t) = \widetilde{U}(t). \tag{5.36}$$

Then the Lebesgue dominated convergence theorem yields

$$\widetilde{U}(t) = \widetilde{U}(0) - \frac{\lambda_1}{\Gamma(\alpha)} \int_0^t (t - s)^{\alpha-1} a(s) \widetilde{U}(s) \, ds + \frac{1}{\Gamma(\alpha)} \int_0^t (t - s)^{\alpha-1} f\left(s, \widetilde{U}(s)\right) ds.$$

Next we claim that F is completely continuous. For that, first we prove that $F : K \to K$ is continuous.

Step 2. Let $\{U_m(t)\}$ be a converging sequence in K to $U(t)$. Then, for any $\epsilon > 0$, let

$$\|U_m(t) - U(t)\| \leq \frac{\Gamma(\alpha)\epsilon}{3\lambda_1 C_1 r_1^{\alpha-\beta}} \left(\frac{\alpha - \beta}{1 - \beta} \right)^{1-\beta}.$$

By Assumption (H2),

$$f(t, U_m(t)) \longrightarrow f(t, U(t))$$

for each $t \in [0, r_1]$ and since

$$\left\| f(t, U_m(t)) - f(t, U(t)) \right\| \leq \frac{\Gamma(\alpha)\epsilon}{3r_1^{\alpha}} \left(\frac{\alpha - \beta}{1 - \beta} \right)^{1-\beta},$$

indeed, we can establish

$$\| FU_m(t) - FU(t) \| \leq \frac{\lambda_1 C_1}{\Gamma(\alpha)} \left(\frac{1-\beta}{\alpha-\beta} \right)^{1-\beta} r_1^{\alpha-\beta} \| U_m(t) - U(t) \|$$

$$+ \frac{r_1^\alpha}{\Gamma(\alpha)} \left(\frac{1-\beta}{\alpha-\beta} \right)^{1-\beta} \| f(t, U_m(t)) - f(t, U(t)) \|$$

$$\leq \epsilon.$$

Taking limit as $m \to \infty$, it is ensured that F is continuous, since ϵ is arbitrary small.

Step 3. Moreover, for $U \in K$,

$$\| FU(t) \| \leq \| U(0) \| + \frac{\lambda_1 C_1 + C_2}{\Gamma(\alpha)} (\| U(0) \| + b) \left(\frac{1-\beta}{\alpha-\beta} \right)^{1-\beta} r_1^{\alpha-\beta}$$

$$\leq \| U(0) \| + b.$$

Hence FK is uniformly bounded. Now it remains to show that F maps K into an equicontinuous family.

Step 4. Now let $U \in K$ and $t_1, t_2 \in J$. Then if $0 < t_1 < t_2 \leq r_1$, by Assumptions (H1)–(H3), we obtain

$$\| FU(t_1) - FU(t_2) \| \leq \frac{\lambda_1}{\Gamma(\alpha)} (\| U(0) \| + b) \int_0^{t_1} \left((t_2 - s)^{\alpha-1} - (t_1 - s)^{\alpha-1} \right) \| a(s) \| \, ds$$

$$+ \frac{\lambda_1}{\Gamma(\alpha)} (\| U(0) \| + b) \int_{t_1}^{t_2} (t_2 - s)^{\alpha-1} \| a(s) \| \, ds$$

$$+ \frac{1}{\Gamma(\alpha)} \left\| \int_0^{t_1} \left((t_2 - s)^{\alpha-1} - (t_1 - s)^{\alpha-1} \right) f(s, U(s)) \, ds \right\|$$

$$+ \frac{1}{\Gamma(\alpha)} \left\| \int_{t_1}^{t_2} (t_2 - s)^{\alpha-1} f(s, U(s)) \, ds \right\|$$

$$\leq \frac{\lambda_1 C_1}{\Gamma(\alpha)} \left(\int_0^{t_1} ((t_2 - s)^{\alpha-1} - (t_1 - s)^{\alpha-1})^{\frac{1}{1-\beta}} \, ds \right)^{1-\beta}$$

$$+ \frac{\lambda_1 C_1}{\Gamma(\alpha)} (\| U(0) \| + b) \left(\int_{t_1}^{t_2} ((t_2 - s)^{\alpha-1})^{\frac{1}{1-\beta}} \, ds \right)^{1-\beta}$$

$$+ \frac{C_2}{\Gamma(\alpha)} (\| U(0) \| + b) \left(\int_0^{t_1} ((t_2 - s)^{\alpha-1} - (t_1 - s)^{\alpha-1})^{\frac{1}{1-\beta}} \, ds \right)^{1-\beta}$$

$$+ \frac{C_2}{\Gamma(\alpha)} (\| U(0) \| + b) \left(\int_{t_1}^{t_2} ((t_2 - s)^{\alpha-1})^{\frac{1}{1-\beta}} \, ds \right)^{1-\beta}.$$

The right-hand side is independent of $U \in K$. Since $0 < \beta < \alpha < 1$, the right-hand side of the above inequality goes to zero as $t_1 \to t_2$. Thus F maps K into an equicontinuous family of functions. In view of Ascoli–Arzela's theorem, F is completely continuous. Then applying the Leray–Schauder fixed point theorem, we deduce that F has a fixed point in K which is a solution of (5.31). $\qquad\square$

5.4.2 Neumann Boundary Condition

Next our aim is to show the existence of solutions of (5.31) with Neumann boundary condition instead of Dirichlet boundary condition. That is,

$$\frac{\partial u(x,t)}{\partial n} = 0, \qquad (x,t) \in \partial\Omega \times J, \tag{5.37}$$

where n is an outward unit normal. Now we define the function $V(t)$ by

$$V(t) = \frac{\int_\Omega u(x,t)\, dx}{\int_\Omega dx} \tag{5.38}$$

and for any constant $b > 0$, take

$$r_2 = \min\left\{ T, \left[\frac{\Gamma(\alpha)b}{(\|V(0)\| + b)C_2} \left(\frac{\alpha - \beta}{1 - \beta} \right)^{1-\beta} \right]^{\frac{1}{\alpha-\beta}} \right\}.$$

Theorem 5.4.2 *Assume that there exists a $\beta \in (0, \alpha)$ for some $\alpha > 0$ such that (H2)–(H3) hold. Then there exists at least one solution for the initial value problem (5.31) on $\Omega \times [0, r_2]$.*

Proof The initial value problem (5.31) with the Neumann boundary condition has a solution if and only if the equation

$$V(t) = V(0) + \frac{1}{\Gamma(\alpha)} \int_0^t (t-s)^{\alpha-1} f(s, V(s))\, ds \tag{5.39}$$

has a solution.

Let $u(x,t)$ be a solution of (5.31). Therefore $u(x,t)$ is a solution of (5.32). Now integrating both sides of (5.32) with respect to $x \in \Omega$ implies

$$\int_\Omega u(x,t)\, dx = \int_\Omega u_0(x)\, dx + \frac{1}{\Gamma(\alpha)} \int_\Omega \int_0^t (t-s)^{\alpha-1} a(s)\Delta u(x,s)\, ds\, dx$$

$$+ \frac{1}{\Gamma(\alpha)} \int_\Omega \int_0^t (t-s)^{\alpha-1} f(s, u(x,s))\, ds\, dx.$$

Using Green's identity (5.4) and the Neumann boundary condition, we obtain

$$\int_\Omega \Delta u(x,t)\, dx = 0.$$

As a consequence, from (5.38) and Assumption (H2), we get

$$V(t) \leq V(0) + \frac{1}{\Gamma(\alpha)} \int_0^t (t-s)^{\alpha-1} f(s, V(s)) \, ds.$$

In order to show the existence of solution of (5.39), define an operator $S : C(J, \mathbb{R}) \to C(J, \mathbb{R})$ by

$$SV(t) = V(0) + \frac{1}{\Gamma(\alpha)} \int_0^t (t-s)^{\alpha-1} f(s, V(s)) \, ds.$$

First we have to prove that the operator S maps K into itself.

$$\left\| SV(t) - V(0) \right\| = \left\| \frac{1}{\Gamma(\alpha)} \int_0^t (t-s)^{\alpha-1} f(s, V(s)) \, ds \right.$$

$$\leq \frac{1}{\Gamma(\alpha)} \int_0^t (t-s)^{\alpha-1} \| f(s, V(s)) \| \, ds.$$

With the help of Assumption (H3) and the Holder inequality, we determine

$$\| SV(t) - V(0) \| \leq \frac{1}{\Gamma(\alpha)} \int_0^t m(s)(t-s)^{\alpha-1} \| V(s) \| ds$$

$$+ \frac{1}{\Gamma(\alpha)} \int_0^t (t-s)^{\alpha-1} \left(\int_0^s m_2(s, \tau) \| V(s) \| \, d\tau \right) ds$$

$$\leq \frac{1}{\Gamma(\alpha)} (\| V(0) \| + b) \left(\int_0^t ((t-s)^{\alpha-1})^{\frac{1}{1-\beta}} \, ds \right)^{1-\beta} \left(\int_0^t (m(s))^{\frac{1}{\beta}} \, ds \right)^{\beta}$$

$$\leq \frac{(\| V(0) \| + b) \, C_2}{\Gamma(\alpha)} \left(\frac{1-\beta}{\alpha - \beta} \right)^{1-\beta} r_2^{\alpha-\beta}$$

$$= \frac{(\| V(0) \| + b) \, C_2}{\Gamma(\alpha)} \left(\frac{1-\beta}{\alpha - \beta} \right)^{1-\beta} r_2^{\alpha-\beta} \leq b.$$

Since K is closed, we next define a sequence $\{V_k(t)\}$ in K which has a subsequence $\{V_{k_i}(t)\}$ such that

$$\lim_{k_i \to \infty} V_{k_i}(t) = \widetilde{V}(t). \tag{5.40}$$

Thus, by Lebesgue's dominated convergence, we obtain

$$\widetilde{V}(t) = \widetilde{V}(0) + \frac{1}{\Gamma(\alpha)} \int_0^t (t-s)^{\alpha-1} f\left(s, \widetilde{V}(s)\right) ds.$$

As in the previous theorem, it is similar to prove that F is completely continuous. Thus, by the Leray–Schauder fixed point theorem, we conclude that S has a fixed point in K which is a solution of (5.31). $\qquad\square$

5.5 Fractional Equations with Kernel

In this section, we consider the fractional partial differential equation with kernel of the form

$$\frac{^{C}\partial^{\alpha}u}{\partial t^{\alpha}} = a(t)\Delta u(x,t) + \int_{0}^{t} h(t-s)\Delta u(x,s)\,\mathrm{d}s + f\big(t,u(x,t)\big), \quad t \in J, \quad (5.41)$$

with the initial condition

$$u(x,0) = u_0(x), \qquad x \in \Omega,$$

where $0 < \alpha < 1, h : J \to \mathbb{R}$ is a positive kernel and $f : J \times \mathbb{R} \to \mathbb{R}$ is a nonlinear function. This Eq. (5.41) is a special case of integrodifferential equation of motion of fractional Maxwell fluid with zero pressure. In order to establish the existence of solution of (5.41), we additionally need the following condition on the kernel along with (H1)–(H3):

$(H4)$ There exists a constant $C_3 > 0$ such that

$$\left(\int_{0}^{T} \left(\int_{0}^{s} h(s-\tau)\,\mathrm{d}\tau \right)^{\frac{1}{\beta}} \mathrm{d}s \right)^{\beta} \leq C_3.$$

The integral equation corresponding to (5.41) can be written as

$$u(x,t) = u_0(x) + \frac{1}{\Gamma(\alpha)} \int_{0}^{t} (t-s)^{\alpha-1} a(s)\Delta u(x,s)\,\mathrm{d}s$$

$$+ \frac{1}{\Gamma(\alpha)} \int_{0}^{t} (t-s)^{\alpha-1} \left(\int_{0}^{s} h(s-\tau)\Delta u(x,\tau)\,\mathrm{d}\tau \right) \mathrm{d}s$$

$$+ \frac{1}{\Gamma(\alpha)} \int_{0}^{t} (t-s)^{\alpha-1} f\big(s,u(x,s)\big)\,\mathrm{d}s. \qquad (5.42)$$

For any constant $b > 0$, take

$$r_3 = \min \left\{ T, \left[\frac{\Gamma(\alpha)b}{(\|U(0)\| + b)(\lambda_1(C_1 + C_3) + C_2)} \left(\frac{\alpha - \beta}{1 - \beta} \right)^{1-\beta} \right]^{\frac{1}{\alpha - \beta}} \right\}.$$

Now we are concerned with the existence of solutions of (5.41) with Dirichlet boundary condition. The main theorem is as follows:

Theorem 5.5.1 *Assume that there exists $\beta \in (0, \alpha)$ for some $0 < \alpha < 1$ such that (H1)–(H3) and (H4) hold. Then there exists at least one solution for the initial value problem (5.41) on $\Omega \times [0, r_3]$.*

Proof Our first aim is to prove that the initial value problem (5.41) has a solution if and only if the equation

$$U(t) = U(0) - \frac{\lambda_1}{\Gamma(\alpha)} \int_0^t (t-s)^{\alpha-1} a(s) U(s) \, ds$$
$$- \frac{\lambda_1}{\Gamma(\alpha)} \int_0^t (t-s)^{\alpha-1} \left(\int_0^s h(s-\tau) U(\tau) \, d\tau \right) ds$$
$$+ \frac{1}{\Gamma(\alpha)} \int_0^t (t-s)^{\alpha-1} f(s, U(s)) \, ds$$

has a solution.

We begin the proof by assuming $u(x, t)$ to be a solution of (5.42). Now multiplying both sides of Eq. (5.42) by $\phi(x)$ and integrating with respect to $x \in \Omega$, we get

$$\int_\Omega \phi(x) u(x, t) \, dx = \int_\Omega \phi(x) u_0(x) \, dx + \frac{1}{\Gamma(\alpha)} \int_\Omega \phi(x) \int_0^t (t-s)^{\alpha-1} a(s) \Delta u(x, s) \, ds \, dx$$
$$+ \frac{1}{\Gamma(\alpha)} \int_\Omega \phi(x) \int_0^t (t-s)^{\alpha-1} \left(\int_0^s h(s-\tau) \Delta u(x, \tau) \, d\tau \right) ds \, dx$$
$$+ \frac{1}{\Gamma(\alpha)} \int_\Omega \phi(x) \int_0^t (t-s)^{\alpha-1} f(s, u(x, s)) \, ds \, dx. \qquad (5.43)$$

Combining (5.35) and Assumptions (H2) and (H4), (5.43) we get

$$U(t) \leq U(0) - \frac{\lambda_1}{\Gamma(\alpha)} \int_0^t (t-s)^{\alpha-1} a(s) U(s) \, ds$$
$$- \frac{\lambda_1}{\Gamma(\alpha)} \int_0^t (t-s)^{\alpha-1} \left(\int_0^s h(s-\tau) U(\tau) \, d\tau \right) ds$$
$$+ \frac{1}{\Gamma(\alpha)} \int_0^t (t-s)^{\alpha-1} f(s, U(s)) \, ds. \qquad (5.44)$$

Let us define an operator $Q : C(J, \mathbb{R}) \to C(J, \mathbb{R})$ by

$$QU(t) = U(0) - \frac{\lambda_1}{\Gamma(\alpha)} \int_0^t (t-s)^{\alpha-1} a(s) U(s) \, ds$$
$$- \frac{\lambda_1}{\Gamma(\alpha)} \int_0^t (t-s)^{\alpha-1} \left(\int_0^s h(s-\tau) U(\tau) \, d\tau \right) ds$$
$$+ \frac{1}{\Gamma(\alpha)} \int_0^t (t-s)^{\alpha-1} f(s, U(s)) \, ds.$$

Next we verify that Q maps K into itself.

$$\| QU(t) - U(0) \| \le \frac{\lambda_1}{\Gamma(\alpha)} (\|U(0)\| + b) \int_0^t (t - s)^{\alpha-1} \|a(s)\| \, ds$$

$$+ \frac{\lambda_1}{\Gamma(\alpha)} (\|U(0)\| + b) \int_0^t (t - s)^{\alpha-1} \left(\int_0^s h(s - \tau) \, d\tau \right) ds$$

$$+ \frac{1}{\Gamma(\alpha)} \int_0^t (t - s)^{\alpha-1} \|f(s, U(s))\| \, ds.$$

Making use of Holder's inequality and the assumptions, for any $U \in K$, we can establish

$$\| QU(t) - U(0) \| \le \frac{\lambda_1 C_1}{\Gamma(\alpha)} (\|U(0)\| + b) \left(\int_0^t \left((t - s)^{\alpha-1} \right)^{\frac{1}{1-\beta}} ds \right)^{1-\beta}$$

$$+ \frac{\lambda_1 C_3}{\Gamma(\alpha)} (\|U(0)\| + b) \left(\int_0^t \left((t - s)^{\alpha-1} \right)^{\frac{1}{1-\beta}} ds \right)^{1-\beta}$$

$$+ \frac{1}{\Gamma(\alpha)} \int_0^t m(s)(t - s)^{\alpha-1} \|U(s)\| \, ds$$

$$\le \frac{(\|U(0)\| + b) \lambda_1 C_1}{\Gamma(\alpha)} \left(\frac{1 - \beta}{\alpha - \beta} \right)^{1-\beta} r_3^{\alpha-\beta}$$

$$+ \frac{(\|U(0)\| + b) \lambda_1 C_3}{\Gamma(\alpha)} \left(\frac{1 - \beta}{\alpha - \beta} \right)^{1-\beta} r_3^{\alpha-\beta}$$

$$+ \frac{(\|U(0)\| + b) C_2}{\Gamma(\alpha)} \left(\frac{1 - \beta}{\alpha - \beta} \right)^{1-\beta} r_3^{\alpha-\beta}$$

$$= \frac{(\|U(0)\| + b) (\lambda_1(C_1 + C_3) + C_2)}{\Gamma(\alpha)} \left(\frac{1 - \beta}{\alpha - \beta} \right)^{1-\beta} r_3^{\alpha-\beta} \le b, \quad t \in [0, r_3].$$

For any sequence $\{U_k(t)\}$ in K, the Lebesgue dominated convergence theorem implies that

$$\tilde{U}(t) = \tilde{U}(0) - \frac{\lambda_1}{\Gamma(\alpha)} \int_0^t (t - s)^{\alpha-1} a(s) \tilde{U}(s) \, ds$$

$$- \frac{\lambda_1}{\Gamma(\alpha)} \int_0^t (t - s)^{\alpha-1} \left(\int_0^s h(s - \tau) \tilde{U}(\tau) \, d\tau \right) ds$$

$$+ \frac{1}{\Gamma(\alpha)} \int_0^t (t - s)^{\alpha-1} f(s, \tilde{U}(s)) \, ds,$$

where $\tilde{U}(t)$ is defined as in (5.36). As in the previous theorem, we can prove that Q is completely continuous. Then, applying the Leray–Schauder fixed point theorem, we achieve that Q has a fixed point in K which is a solution of (5.41). $\qquad\square$

The following theorem asserts the existence of solution of (5.41) with Neumann boundary condition (5.37).

Theorem 5.5.2 *Assume that there exists $\beta \in (0, \alpha)$ for some $0 < \alpha < 1$ such that (H2) and (H3) hold. Then there exists at least one solution for the initial value problem (5.41) on $\Omega \times [0, r_2]$.*

Proof In order to prove the existence of solutions of (5.41), it is enough to show that the equation

$$V(t) = V(0) + \frac{1}{\Gamma(\alpha)} \int_0^t (t - s)^{\alpha-1} f(s, V(s)) \, ds$$

has a solution.

Assume $u(x, t)$ to be a solution of (5.41). Then it follows that $u(x, t)$ is a solution of (5.42). Now integrating both sides of Eq. (5.42) with respect to $x \in \Omega$, we are led to

$$\int_\Omega u(x, t) \, dx = \int_\Omega u_0(x) \, dx + \frac{1}{\Gamma(\alpha)} \int_\Omega \int_0^t (t - s)^{\alpha-1} a(s) \Delta u(x, s) \, ds \, dx$$
$$+ \frac{1}{\Gamma(\alpha)} \int_\Omega \int_0^t (t - s)^{\alpha-1} \left(\int_0^s h(s - \tau) \Delta u(x, \tau) \, d\tau \right) ds \, dx$$
$$+ \frac{1}{\Gamma(\alpha)} \int_\Omega \int_0^t (t - s)^{\alpha-1} f(s, u(x, s)) \, ds \, dx. \tag{5.45}$$

Combining Green's identity, the Neumann boundary condition and Assumption (H2), (5.45) can be written as

$$V(t) \le V(0) + \frac{1}{\Gamma(\alpha)} \int_0^t (t - s)^{\alpha-1} f(s, V(s)) \, ds.$$

Now an operator $P : C(J, \mathbb{R}) \to C(J, \mathbb{R})$ is defined by

$$PV(t) = V(0) + \frac{1}{\Gamma(\alpha)} \int_0^t (t - s)^{\alpha-1} f(s, V(s)) \, ds.$$

Next we have to prove that the operator P maps K into itself. From the above equation, we observe that

$$\| PV(t) - V(0) \| \le \frac{1}{\Gamma(\alpha)} \int_0^t (t - s)^{\alpha-1} \| f(s, V(s)) \| \, ds \le b.$$

Since K is closed, for any sequence $\{V_k(t)\}$ in K and $\widetilde{V}(t)$ as in (5.40), the Lebesgue dominated convergence theorem gives

$$\tilde{V}(t) = \tilde{V}(0) + \frac{1}{\Gamma(\alpha)} \int_0^t (t-s)^{\alpha-1} f(s, \tilde{V}(s)) \, ds.$$

It is easy to show that P is completely continuous. Then, by the Leray–Schauder fixed point theorem, we conclude that P has a fixed point in K which is a solution of (5.41). □

5.6 Examples

Example 5.6.1 Consider the following nonlinear fractional partial differential equation:

$$\frac{{}^C\partial^{\frac{1}{2}} u(x,t)}{\partial t^{\frac{1}{2}}} = t^2 \Delta u(x,t) + u^2(x,t), \quad (x,t) \in \Omega \times J \tag{5.46}$$

with the initial condition

$$u(x,0) = u_0, \ x \in \Omega$$

and the boundary condition

$$u(x,t) = 0, \ (x,t) \in \partial\Omega \times J,$$

where $J = [0,1]$ and $\Omega = [0, \pi/2]$. Here $a(t) = t^2$, and

$$f(t, u(x,t)) = u^2(x,t).$$

Since the eigenfunctions of the Laplacian operator are $\sin mx$ and $\cos mx$ where $\lambda = m^2$, we note that Assumptions (H1)–(H3) of Theorem 5.4.1 are satisfied for some $\beta \in (0, 1/2)$. Hence Problem (5.46) has a solution.

Example 5.6.2 Consider the fractional partial differential equation with kernel of the form

$$\frac{{}^C\partial^{\frac{1}{2}} u(x,t)}{\partial t^{\frac{1}{2}}} = \Delta u(x,t) + \int_0^t e^{-(t-s)} \Delta u(x,s) \, ds + u^2(x,t), \quad (x,t) \in \Omega \times J \tag{5.47}$$

with the initial condition

$$u(x,0) = u_0, \ x \in \Omega$$

and the boundary condition

$$u(x,t) = 0, \ (x,t) \in \partial\Omega \times J,$$

where $J = [0, 1]$ and $\Omega = [0, \pi/2]$. Here $a(t) = 1$, $h(t) = e^{-t}$, and

$$f(t, u(x, t)) = u^2(x, t).$$

Note that there is a constant $K > 0$ such that

$$\left(\int_0^1 \left(\int_0^s e^{-(t-s)} \, dt \right)^{\frac{1}{\beta}} ds \right)^{\beta} \le K.$$

Observe that Assumption (H4) is satisfied in addition to Assumptions (H1)–(H3) of Theorem 5.5.1 for some $\beta \in (0, 1/2)$. Hence Eq. (5.47) has a solution.

5.7 Exercises

5.1. Solve the heat equation

$$\frac{\partial u(x, t)}{\partial t} = \frac{\partial^2 u(x, t)}{\partial x^2}, 0 \le x \le l, \ t > 0,$$

with initial condition $u(x, 0) = f(x)$.

5.2. Solve the wave equation

$$\frac{\partial^2 u(x, t)}{\partial t^2} = \frac{\partial^2 u(x, t)}{\partial x^2}, 0 \le x \le l, \ t > 0,$$

with initial condition $u(x, 0) = f(x)$.

5.3. Using the Adomian decomposition method (ADM), solve the following fractional partial differential equation:

$$\frac{{}^C\partial^\alpha u(x, t)}{\partial t^\alpha} + \frac{{}^C\partial^\beta u(x, t)}{\partial t^\beta} + u(x, t) = \frac{\partial^2 u(x, t)}{\partial x^2},$$

$$0 \le x \le l, \ t > 0, \ 1 < \alpha \le 2, \ \frac{1}{2} < \beta \le 1,$$

with initial conditions $u(x, 0) = 0$, $u_t(x, 0) = e^x$.

5.4. Solve the equation by the ADM

$$\frac{{}^C\partial^\alpha u(x, t)}{\partial t^\alpha} + \frac{{}^C\partial^{\alpha-1} u(x, t)}{\partial t^{\alpha-1}} + u(x, t) = \frac{\partial^2 u(x, t)}{\partial x^2} + \sinh u,$$

$$0 \le x \le l, \ t > 0, \ 1 < \alpha \le 2,$$

with initial conditions $u(x, 0) = u_t(x, 0) = 0$.

5.5. Solve the equation by the ADM

$$\frac{{}^C\partial^\alpha u(x,t)}{\partial t^\alpha} - \frac{\partial u(x,t)}{\partial x} = \frac{\partial^2 u(x,t)}{\partial x^2} + \frac{3t^2}{2}\sinh x,$$
$$0 \le x \le l, \ t > 0, \ 0 < \alpha \le 1,$$

with initial conditions $u(x,0) = u_t(x,0) = 0$.

5.6. Solve the equation by the ADM

$$\frac{{}^C\partial^\alpha u(x,t)}{\partial t^\alpha} - x\frac{\partial u(x,t)}{\partial x} = \frac{\partial}{\partial x}\left((x^2+3)\frac{\partial u(x,t)}{\partial x}\right) + xe^t + 1,$$
$$0 \le x \le l, \ t > 0, \ 0 < \alpha \le 1,$$

with initial conditions $u(x,0) = 1 + x$, $u_t(x,0) = 2x$.

5.7. Solve

$$\frac{{}^C\partial^\alpha u(x,t)}{\partial t^\alpha} = \frac{\partial^2 u(x,t)}{\partial x^2}$$
$$0 \le x \le l, \ t > 0, \ 0 < \alpha \le 1,$$

with initial condition $u(x,0) = f(x)$.

5.8. Solve

$$i\frac{{}^C\partial^\alpha \psi(x,t)}{\partial t^\alpha} + \frac{1}{2}\frac{{}^C\partial^\beta}{\partial x^\beta}\left(\frac{{}^C\partial^\beta \psi(x,t)}{\partial x^\beta}\right) = 0,$$
$$0 \le x \le l, \ t > 0, \ 0 < \alpha, \beta \le 1,$$

with initial conditions $\psi(x,0) = \psi_0(x)$.

5.9. Establish the existence of solution of the nonlinear equation

$$\frac{{}^C\partial^{\frac{1}{2}}u}{\partial t^{\frac{1}{2}}} = \Delta u(x,t) + \sin u(x,t)$$
$$u(x,0) = u_0(x).$$

References

1. Nigmatullin, R.R.: The realization of the generalized transfer equation in a medium with fractal geometry. Physica Status Solidi B **133**, 425–430 (1986)
2. Chen, P., Ma, W., Tao, S., Zhang, K.: Blowup and global existence of mild solutions for fractional extended Fisher-Kolmogorov equations. Int. J. Nonlinear Sci. Numer. Simul. **22**, 641–656 (2021)

3. Langlands, T.A.M., Henry, B.I., Wearne, S.L.: Fractional cable equation models for anomalous electrodiffusion in nerve cells: infinite domain solutions. SIAM J. Appl. Math. **71**, 1168–1203 (2011)
4. Hall, M.G., Barrick, T.R.: Two step anomalous diffusion tensor imaging. NMR Biomed. **25**, 286–294 (2012)
5. Biler, P., Funaki, T., Woyczynski, W.: Fractal Burgers equations. J. Diff. Equ. **148**, 9–46 (1998)
6. Momani, S.: Non-perturbative analytical solutions of the space and time fractional Burgers equations. Chaos Solitons Fractals **28**, 930–937 (2006)
7. Gurney, W.S.C., Nisbet, R.M.: The regulation of inhomogeneous populations. J. Theor. Biol. **54**, 35–49 (1977)
8. Guo, B., Pu, X., Huang, F.: Fractional Partial Differential Equations and their Numerical Solutions. Science Press, Beijing (2011)
9. Kubica, A., Ryszewska, K., Yamamoto, M.: Time-Fractional Differential Equations; A Theoretical Introduction. Springer, Singapore (2020)
10. Joice Nirmala, R., Balachandran, K.: Analysis of solutions of time fractional telegraph equation. J. Korean Soc. Ind. Appl. Math. **14**, 209–224 (2014)
11. Kolokoltsov, V.: The probabilistic point of view on the generalized fractional partial differential equations. Fract. Calc. Appl. Anal. **22**, 543–600 (2019)
12. Kilbas, A.A., Srivastava, H.M., Trujillo, J.J.: Theory and Applications of Fractional Differential Equations. Elsevier, Amsterdam (2006)
13. Evans, L.C.: Partial Differential Equations. American Mathematical Society, Providence (1998)
14. Adomian, G.: Solving Frontier Problems of Physics: The Decomposition Method. Kluwer Academic Publishers, Boston, MA (1994)
15. Gülkac, V.: The homotopy perturbation method for the Black-Scholes equation. J. Stat. Comput. Simul. **80**, 1349–1354 (2010)
16. Kumar, S., Yildirim, A., Khan, Y., Jafari, H., Sayevand, K., Wei, L.: Analytical solution of fractional Black-Scholes European option pricing equation by using Laplace transform. J. Fract. Calc. Appl. **2**, 1–9 (2012)
17. Ouyang, Z.: Existence and uniqueness of the solutions for a class of nonlinear fractional order partial differential equations with delay. Comput. Math. Appl. **61**, 860–870 (2011)

Chapter 6
Fractional Integrals and Derivatives

Abstract In this chapter, we list various types of fractional integrals and fractional derivatives available in the literature. In fact, the purpose is to show that there are many types of fractional integrals and derivatives currently under investigation by several researchers some with theory and others with applications. A brief comment about few fractional derivatives is given. Several examples and exercises are constructed to understand different definitions.

Keywords Definitions · Fractional integrals · Fractional derivatives

In this chapter, we list various types of fractional integrals and fractional derivatives available in the literature for the benefit of the readers. These definitions and their properties can be found in [1–7] and in particular [8, 9]. In fact, the purpose of this chapter is to show that there are many types of fractional integrals and derivatives currently under investigation by several researchers some with theory and others with applications. This may provide an insight to the students about the developments of the field. Here the functions are taken in appropriate function spaces with required smoothness conditions.

6.1 Definitions of Fractional Integrals

1. Liouville left-sided integral [8]:

$$I_+^\alpha f(x) = \frac{1}{\Gamma(\alpha)} \int_{-\infty}^x (x - \xi)^{\alpha-1} f(\xi) d\xi. \tag{6.1}$$

2. Liouville right-sided integral [8]:

$$I_-^\alpha f(x) = \frac{1}{\Gamma(\alpha)} \int_x^\infty (\xi - x)^{\alpha-1} f(\xi) d\xi. \tag{6.2}$$

3. Riemann–Liouville left-sided integral [8]:

$$I_{a+}^{\alpha} f(x) = \frac{1}{\Gamma(\alpha)} \int_a^x (x - \xi)^{\alpha-1} f(\xi) d\xi, \quad x \geq a. \tag{6.3}$$

4. Riemann–Liouville right-sided integral [8]:

$$I_{b-}^{\alpha} f(x) = \frac{1}{\Gamma(\alpha)} \int_x^b (\xi - x)^{\alpha-1} f(\xi) d\xi, \quad x \leq b. \tag{6.4}$$

5. Hadamard integral [8]:

$$I_+^{\alpha} f(x) = \frac{1}{\Gamma(\alpha)} \int_0^x \frac{f(\xi)}{(ln(\frac{x}{\xi}))^{1-\alpha}} \frac{d\xi}{\xi}, \quad x > 0, \alpha > 0. \tag{6.5}$$

6. Weyl integral [4]:

$$_xW_{\infty}^{\alpha} f(x) = \frac{1}{\Gamma(\alpha)} \int_x^{\infty} (\xi - x)^{\alpha-1} f(\xi) d\xi. \tag{6.6}$$

7. Erdelyi left-sided integral [8]:

$$I_{\sigma,\eta}^{\alpha} f(x) = \frac{\sigma x^{-\sigma(\alpha+\eta)}}{\Gamma(\alpha)} \int_{-\infty}^x (x^{\sigma} - \xi^{\sigma})^{\alpha-1} \xi^{\sigma\eta+\sigma-1} f(\xi) d\xi \tag{6.7}$$

8. Erdelyi right-sided integral [8]:

$$I_{\sigma,\eta}^{\alpha} f(x) = \frac{\sigma x^{\sigma\eta}}{\Gamma(\alpha)} \int_x^{\infty} (\xi^{\sigma} - x^{\sigma})^{\alpha-1} \xi^{\sigma(1-\alpha-\eta)-1} f(\xi) d\xi. \tag{6.8}$$

9. Kober left-sided integral [8]:

$$I_{1,\eta}^{\alpha} f(x) = \frac{x^{-\alpha-\eta}}{\Gamma(\alpha)} \int_0^x (x - \xi)^{\alpha-1} \xi^{\eta} f(\xi) d\xi. \tag{6.9}$$

10. Kober right-sided integral [8]:

$$I_{1,\eta}^{\alpha} f(x) = \frac{x^{\eta}}{\Gamma(\alpha)} \int_x^{\infty} (\xi - x)^{\alpha-1} \xi^{-\alpha-\eta} f(\xi) d\xi. \tag{6.10}$$

11. Riemann–Liouville left integral of variable fractional order [8]:

$$_aI_x^{\alpha(\cdot,\cdot)} f(x) = \int_a^x (\xi - x)^{\alpha(\xi,x)-1} f(\xi) \frac{d\xi}{\Gamma(\alpha(\xi,x))}. \tag{6.11}$$

12. Riemann–Liouville right integral of variable fractional order [8]:

$$_xI_b^{\alpha(\cdot,\cdot)}f(x) = \int_x^b (x-\xi)^{\alpha(\xi,x)-1} f(\xi)\frac{d\xi}{\Gamma(\alpha(\xi,x))}. \tag{6.12}$$

13. Generalized left-sided integral [10–12]:

$$^\rho I_{a+}^\alpha f(x) = \frac{\rho^{1-\alpha}}{\Gamma(\alpha)} \int_a^x \frac{s^{\rho-1} f(s)}{(x^\rho - s^\rho)^{1-\alpha}} ds \tag{6.13}$$

14. Generalized right-sided integral [10–12]:

$$^\rho I_{b-}^\alpha f(x) = \frac{\rho^{1-\alpha}}{\Gamma(\alpha)} \int_x^b \frac{s^{\rho-1} f(s)}{(s^\rho - x^\rho)^{1-\alpha}} ds \tag{6.14}$$

15. Fractional integral of a function f with respect to a function g of order α [9]

$$I_{a+,g}^\alpha f(x) = \frac{1}{\Gamma(\alpha)} \int_a^x \frac{f(t)}{[g(x)-g(t)]^{1-\alpha}} g'(t)dt, \ \alpha > 0, \ a < b, \tag{6.15}$$

for every function $f \in L_1(a,b)$ and for any monotonic function $g(t)$ with continuous derivative.

6.2 Definitions of Fractional Derivatives

1. Liouville derivative, for $0 < \alpha < 1$, [8]:

$$D^\alpha f(x) = \frac{1}{\Gamma(1-\alpha)} \frac{d}{dx} \int_{-\infty}^x (x-\xi)^{-\alpha} f(\xi)d\xi, \quad -\infty < x < +\infty. \tag{6.16}$$

2. Liouville left-sided derivative, for $n - 1 < \alpha < n$, [8]:

$$D_+^\alpha f(x) = \frac{1}{\Gamma(n-\alpha)} \frac{d^n}{dx^n} \int_{-\infty}^x (x-\xi)^{-\alpha+n-1} f(\xi)d\xi, \quad -\infty < x. \tag{6.17}$$

3. Liouville right-sided derivative for $n - 1 < \alpha < n$ [8]:

$$D_-^\alpha f(x) = \frac{(-1)^n}{\Gamma(n-\alpha)} \frac{d^n}{dx^n} \int_x^\infty (x-\xi)^{-\alpha+n-1} f(\xi)d\xi, \quad x < \infty. \tag{6.18}$$

4. Riemann–Liouville left-sided derivative [8]:

$$D_{a+}^\alpha f(x) = \frac{1}{\Gamma(n-\alpha)} \frac{d^n}{dx^n} \int_a^x (x-\xi)^{n-\alpha-1} f(\xi)d\xi, \quad x \geq a. \tag{6.19}$$

5. Riemann–Liouville right-sided derivative [8]:

$$D_{b-}^{\alpha} f(x) = \frac{(-1)^n}{\Gamma(n-\alpha)} \frac{d^n}{dx^n} \int_x^b (\xi - x)^{n-\alpha-1} f(\xi) d\xi, \quad x \le b. \quad (6.20)$$

6. Caputo left-sided derivative [13]:

$$^C D_{a+}^{\alpha} f(x) = \frac{1}{\Gamma(n-\alpha)} \int_a^x (x - \xi)^{n-\alpha-1} f^{(n)}(\xi) d\xi, \quad x \ge a. \quad (6.21)$$

7. Caputo right-sided derivative [13]:

$$^C D_{b-}^{\alpha} f(x) = \frac{(-1)^n}{\Gamma(n-\alpha)} \int_x^b (\xi - x)^{n-\alpha-1} f^{(n)}(\xi) d\xi, \quad x \le b. \quad (6.22)$$

8. Grunwald–Letnikov left-sided derivative [5]:

$$^{GL} D_{a+}^{\alpha} f(x) = \lim_{h \to 0} \frac{1}{h^\alpha} \sum_{k=0}^{[n]} (-1)^k \frac{\Gamma(\alpha+1) f(x-kh)}{\Gamma(k+1)\Gamma(\alpha-k+1)}, \quad nh = x - a. \quad (6.23)$$

9. Grunwald–Letnikov right-sided derivative [5]:

$$^{GL} D_{b-}^{\alpha} f(x) = \lim_{h \to 0} \frac{1}{h^\alpha} \sum_{k=0}^{[n]} (-1)^k \frac{\Gamma(\alpha+1) f(x+kh)}{\Gamma(k+1)\Gamma(\alpha-k+1)}, \quad nh = b - x. \quad (6.24)$$

10. Weyl derivative [4]:

$$_x D_{\infty}^{a} f(x) = \frac{(-1)^n}{\Gamma(\alpha)} \frac{d^n}{dx^n} \int_x^{\infty} (\xi - x)^{\alpha-1} f(\xi) d\xi. \quad (6.25)$$

11. Marchaud derivative [9]:

$$D_+^{\alpha} f(x) = \frac{\alpha}{\Gamma(1-\alpha)} \int_{-\infty}^x \frac{f(x) - f(\xi)}{(x-\xi)^{1+\alpha}} d\xi. \quad (6.26)$$

12. Marchaud left-sided derivative [9]:

$$D_+^{\alpha} f(x) = \frac{\alpha}{\Gamma(1-\alpha)} \int_0^{\infty} \frac{f(x) - f(x-\xi)}{\xi^{1+\alpha}} d\xi. \quad (6.27)$$

13. Marchaud right-sided derivative [9]:

$$D_-^{\alpha} f(x) = \frac{\alpha}{\Gamma(1-\alpha)} \int_0^{\infty} \frac{f(x) - f(x+\xi)}{\xi^{1+\alpha}} d\xi. \quad (6.28)$$

14. Hadamard derivative [8]:

$$D^\alpha_+ f(x) = \frac{\alpha}{\Gamma(1-\alpha)} \int_0^x \frac{f(x) - f(\xi)}{(ln(\frac{x}{\xi}))^{1+\alpha}} \frac{d\xi}{\xi}. \tag{6.29}$$

15. Davison–Essex derivative [14]:

$$D^{n+\alpha}_0 f(x) = \frac{1}{\Gamma(1-\alpha)} \frac{d^{n+1-k}}{dx^{n+1-k}} \int_0^x (x - \xi)^{-\alpha} f^{(k)}(\xi) d\xi. \tag{6.30}$$

16. Riesz derivative, for $n - 1 < \alpha < n$, [8]:

$$D^\alpha_x f(x) = -\frac{1}{2cos(\alpha\pi/2)\Gamma(\alpha)} \frac{d^n}{dx^n} \left\{ \int_{-\infty}^x (x - \xi)^{n-\alpha-1} f(\xi) d\xi \right.$$
$$\left. + \int_x^\infty (\xi - x)^{n-\alpha-1} f(\xi) d\xi \right\}. \tag{6.31}$$

17. Coimbra derivative [15]:

$$D^{\alpha(x)}_0 f(x) = \frac{1}{\Gamma(1-\alpha(x))} \left\{ \int_0^x (x - \xi)^{-\alpha(x)} f'(\xi) d\xi + f(0)x^{-\alpha(x)} \right\} \tag{6.32}$$

18. Cossar derivative [7]:

$$D^\alpha_- f(x) = -\frac{1}{\Gamma(1-\alpha)} \lim_{N\to\infty} \frac{d}{dx} \int_x^N (\xi - x)^{-\alpha} f(\xi) d\xi. \tag{6.33}$$

19. Riemann–Liouville left derivative of variable fractional order [16]:

$$_a D^{\alpha(\cdot,\cdot)}_x f(x) = \frac{d}{dx} \int_a^x (x - \xi)^{-\alpha(\xi,x)} f(\xi) \frac{d\xi}{\Gamma(1 - \alpha(\xi, x))}. \tag{6.34}$$

20. Riemann–Liouville right derivative of variable fractional order [16]:

$$_x D^{\alpha(\cdot,\cdot)}_b f(x) = \frac{d}{dx} \int_x^b (\xi - x)^{-\alpha(\xi,x)} f(\xi) \frac{d\xi}{\Gamma(1 - \alpha(\xi, x))}. \tag{6.35}$$

21. Caputo left derivative of variable fractional order [8]:

$$^C_a D^{\alpha(\cdot,\cdot)}_x f(x) = \int_a^x (x - \xi)^{-\alpha(\xi,x)} f'(\xi) \frac{d\xi}{\Gamma(1 - \alpha(\xi, x))}. \tag{6.36}$$

22. Caputo right derivative of variable fractional order [8]:

$$_x^C D_b^{\alpha(\cdot,\cdot)} f(x) = \int_x^b (\xi - x)^{-\alpha(\xi,x)} f'(\xi) \frac{d\xi}{\Gamma(1 - \alpha(\xi, x))}. \qquad (6.37)$$

23. Caputo derivative of variable fractional order [8]:

$$^C D_x^{\alpha(x)} f(x) = \frac{1}{\Gamma(1 - \alpha(x))} \int_0^x (x - \xi)^{-\alpha(\xi,x)} f'(\xi) d\xi. \qquad (6.38)$$

24. Modified Riemann–Liouville fractional derivative [8]:

$$D^\alpha f(x) = \frac{1}{\Gamma(1 - \alpha)} \frac{d}{dx} \int_0^x (x - \xi)^{-\alpha} [f(\xi) - f(0)] d\xi. \qquad (6.39)$$

25. Osler fractional derivative [17]:

$$_a D_z^\alpha f(z) = \frac{\Gamma(\alpha + 1)}{2\pi i} \int_C \frac{f(\xi)}{(\xi - z)^{1+\alpha}} d\xi, \qquad (6.40)$$

where C is the contour integral.

26. Hilfer derivative [18]:

$$D^{\mu,\gamma} f(x) = I^{\gamma(1-\mu)} \frac{d}{dx} I^{(1-\mu)(1-\gamma)} f(x). \qquad (6.41)$$

27. Caputo–Fabrizio fractional derivative of order $0 < \alpha < 1$ [19]:

$$^{CF} D^\alpha f(t) = \frac{M(\alpha)}{(1 - \alpha)} \int_a^t \exp\left[-\frac{\alpha(t - \tau)}{1 - \alpha} \right] f'(\tau) d\tau, \qquad (6.42)$$

where $M(\alpha)$ is a normalization function with $M(0) = M(1) = 1$.

28. Atangana–Baleanu fractional derivative in Riemann–Liouville sense [20]:

$$^{ABR}_a D_t^\alpha f(t) = \frac{AB(\alpha)}{1 - \alpha} \frac{d}{dt} \int_a^t E_\alpha \left(\frac{-\alpha(t - \tau)^\alpha}{1 - \alpha} \right) f(\tau) d\tau \qquad (6.43)$$

where $AB(\alpha)$ is a normalization function with $AB(0) = AB(1) = 1$.

29. Atangana–Baleanu derivative in Caputo sense [20]:

$$^{ABC}_a D_t^\alpha f(t) = \frac{AB(\alpha)}{1 - \alpha} \int_a^t E_\alpha \left(\frac{-\alpha(t - \tau)^\alpha}{1 - \alpha} \right) f'(\tau) d\tau. \qquad (6.44)$$

30. Generalized left-sided derivative (RL sense) [12]:

$$D_{a+}^{\alpha,\rho} f(x) = \frac{\rho^{\alpha-n+1}}{\Gamma(n - \alpha)} \left(x^{1-\rho} \frac{d}{dx} \right)^n \int_a^x \frac{s^{\rho-1} f(s)}{(x^\rho - s^\rho)^{\alpha-n+1}} ds \qquad (6.45)$$

31. Generalized right-sided derivative (RL sense) [12]:

$$D_{b^-}^{\alpha,\rho} f(x) = \frac{\rho^{\alpha-n+1}}{\Gamma(n-\alpha)} \left(-x^{1-\rho}\frac{d}{dx}\right)^n \int_x^b \frac{s^{\rho-1} f(s)}{(s^\rho - x^\rho)^{\alpha-n+1}} ds \quad (6.46)$$

32. Generalized left-sided derivative (Caputo sense) [12]:

$$^C D_{a^+}^{\alpha,\rho} f(x) = \frac{\rho^{\alpha-n+1}}{\Gamma(n-\alpha)} \int_a^x \frac{s^{(\rho-1)(1-n)}}{(x^\rho - s^\rho)^{\alpha-n+1}} f^n(s) ds \quad (6.47)$$

33. Generalized right-sided derivative (Caputo sense) [12]:

$$^C D_{b^-}^{\alpha,\rho} f(x) = \frac{(-1)^n \rho^{\alpha-n+1}}{\Gamma(n-\alpha)} \int_x^b \frac{s^{(\rho-1)(1-n)}}{(s^\rho - x^\rho)^{\alpha-n+1}} f^n(s) ds \quad (6.48)$$

34. R-L derivative of a function f with respect to a function g of order $0 < \alpha < 1$ [9]

$$D_g^\alpha f(x) = \frac{1}{\Gamma(1-\alpha)} \frac{1}{g'(x)} \frac{d}{dx} \int_a^x \frac{f(t)}{[g(x) - g(t)]^\alpha} g'(t) dt \quad (6.49)$$

35. Marchaud derivative of a function f with respect to a function g of order $0 < \alpha < 1$ [9]

$$D_{a^+,g}^\alpha f(x) = \frac{1}{\Gamma(1-\alpha)} \frac{f(x)}{[g(x) - g(a)]^\alpha}$$
$$+ \frac{\alpha}{\Gamma(1-\alpha)} \int_a^x \frac{f(x) - f(t)}{[g(x) - g(t)]^{1+\alpha}} g'(t) dt \quad (6.50)$$

The above definitions are not complete and many more new definitions will crop up as and when new problems arise in different fields of research.

6.3 Comments

Recently, it has been established that the fractional derivative introduced via regular kernel has no physical significance and those derivatives contain nothing new [21–25]. In fact, they are not satisfying the rigorous analysis of fractional derivative [26, 27].

Several authors have introduced new definitions of fractional derivatives, such as conformable fractional derivative, deformable derivative, α-derivative, M-fractional derivative, and so on. In fact, these concepts do not bring any novelty and have no physical relevance, they also create a certain confusion by using the term fractional and have nothing to do with the classical Riemann–Liouville fractional derivative.

Moreover, analyzing the definition of these derivatives, it is easy to observe that one can introduce a new fractional derivative without any physical significance.

(i) Conformal derivative [28]:

$$D^\alpha f(t) = \lim_{h \to 0} \frac{f(t + ht^{1-\alpha}) - f(t)}{h}. \tag{6.51}$$

(ii) α-derivative (see [29]):

$$D^\alpha f(t) = \lim_{h \to 0} \frac{f(te^{ht^{-\alpha}}) - f(t)}{h}. \tag{6.52}$$

(iii) Local M-derivative [29]:

$$D_M^{\alpha,\beta} f(t) = \lim_{h \to 0} \frac{f(t E_\beta(ht^{-\alpha})) - f(t)}{h} \tag{6.53}$$

where $E_\beta(.)$ is the Mittag–Leffler function. If f is differentiable, then

$$D_M^{\alpha,\beta} f(t) = \frac{t^{1-\alpha}}{\Gamma(\beta + 1)} \frac{df}{dt}$$

and $\beta = 1$ it is equal to α-derivative. Also $D_M^{\alpha,\beta}({}_M I_a^{\alpha,\beta} f(t)) = f(t)$, where

$$_M I_a^{\alpha,\beta} f(t) = \Gamma(1 + \beta) \int_a^t \frac{f(s)}{s^{1-\alpha}} ds. \tag{6.54}$$

Now let us introduce a general definition. Let ϕ be a strictly positive function defined on an interval $J \subset \mathbb{R}_+$ and let $f : (0, \infty) \to \mathbb{R}$ be a given function. Then we define the "ϕ-fractional derivative" of f by

$$D^{(\phi)} f(t) = \lim_{\epsilon \to 0} \frac{f(t + \epsilon\phi(t)) - f(t)}{\epsilon}, \quad t > 0, \tag{6.55}$$

provided that the limit exists. It is easy to see that from

$$\lim_{\epsilon \to 0} \frac{f(t + \epsilon\phi(t)) - f(t)}{\epsilon} = \lim_{\epsilon \to 0} \frac{f(t + \epsilon\phi(t)) - f(t)}{\epsilon\phi(t)} \phi(t) = f'(t)\phi(t),$$

it follows that $D^{(\phi)} f(t) = f'(t)\phi(t)$, $t > 0$, that is, $D^{(\phi)} f(t)$ represents the derivative of a given function $f(t)$ with respect to another function $\phi(t)$, a concept that is well known even for fractional derivatives. Next, if we take $\phi(t) = t^{1-\alpha}$ for $t > 0$ and $\alpha \in (0, 1)$, then from (6.55), we obtain the definition of the "conformable fractional derivative" of f of order α as it is defined in [28]. Next, for each $\epsilon > 0$, let us take

$$\phi_\epsilon(t) = t^{1-\alpha} \sum_{k=0}^{\infty} \frac{\epsilon^k t^{-k\alpha}}{(k+1)!}$$

for $t > 0$ and $\alpha \in (0, 1)$. Then it is easy to see that

$$t e^{\epsilon t^{-\alpha}} = t + \epsilon \phi_\epsilon(t)$$

so that the α fractional derivative can be defined by (6.55) because

$$D^\alpha f(t) = \lim_{\epsilon \to 0} \frac{f(t e^{\epsilon t^{-\alpha}}) - f(t)}{\epsilon} = \lim_{\epsilon \to 0} \frac{f(t + \epsilon \phi(t)) - f(t)}{\epsilon}$$
$$= \lim_{\epsilon \to 0} D^{(\phi_\epsilon)} f(t) = f'(t) \lim_{\epsilon \to 0} \phi_\epsilon(t),$$

provided that the limit exists. Finally, if we take

$$\phi_\epsilon(t) = t^{1-\alpha} \sum_{k=0}^{\infty} \frac{\epsilon^k t^{-k\alpha}}{(\alpha k + 1)!}$$

for $t > 0$ and $\alpha \in (0, 1)$, then

$$t E_\alpha(\epsilon t^{-\alpha}) = t + \epsilon \phi_\epsilon(t)$$

and so the M-fractional derivative $M^\alpha f$, defined in [29], can be obtained by (6.55).

Similarly by using the Mittag–Leffler functions with two parameters, three parameters, etc, one can define a new derivative, or instead of ϕ, one can take any analytic function and introduce a new kind of fractional derivative without any purpose.

6.4 Examples

Example 6.4.1 Let us evaluate the Hadamard fractional integral of the function $f(x) = (\ln(\frac{x}{a}))^{\beta-1}$.

From the definition of Hadamard integral, we have

$$I_{a+}^\alpha f(x) = I_{a+}^\alpha (\ln(\frac{x}{a}))^{\beta-1} = \frac{1}{\Gamma(\alpha)} \int_a^x (\ln \frac{x}{a})^{\alpha-1} (\ln \frac{t}{a})^{\beta-1} \frac{dt}{t}.$$

Introducing the change of variable $\eta = (\ln \frac{t}{a})/(\ln \frac{x}{a})$, the limits become $\eta = 0$ when $t = a$ and $\eta = 1$ when $t = x$ and $(\ln \frac{x}{a})d\eta = \frac{dt}{t}$. Substituting this in the above integral, we get

$$I_{a+}^{\alpha}(\ln(\frac{x}{a}))^{\beta-1} = \frac{1}{\Gamma(\alpha)}(\ln\frac{x}{a})^{\alpha+\beta-1}\int_0^1(1-\eta)^{\alpha-1}\eta^{\beta-1}d\eta.$$

Since $\int_0^1(1-\eta)^{\alpha-1}\eta^{\beta-1}d\eta = \frac{\Gamma(\alpha)\Gamma(\beta)}{\Gamma(\alpha+\beta)}$, we have

$$I_{a+}^{\alpha}(\ln(\frac{x}{a}))^{\beta-1} = \frac{\Gamma(\beta)}{\Gamma(\alpha+\beta)}(\ln\frac{x}{a})^{\alpha+\beta-1}.$$

Example 6.4.2 Property of Weyl integral:

$$W_x^{\alpha}W_x^{\beta}f(x) = W_x^{\alpha+\beta}f(x), x > 0, \alpha > 0, \beta > 0.$$

From the definition $W_x^{\alpha}f(x) = \frac{1}{\Gamma(\alpha)}\int_x^{\infty}(t-x)^{\alpha-1}f(t)dt$ and so,

$$W_x^{\beta}W_x^{\alpha}f(x) = \frac{1}{\Gamma(\alpha)}W_x^{\beta}\left[\int_x^{\infty}(t-x)^{\alpha-1}f(t)dt\right],$$
$$= \frac{1}{\Gamma(\alpha)}\frac{1}{\Gamma(\beta)}\int_x^{\infty}(s-x)^{\beta-1}ds\left[\int_s^{\infty}(t-s)^{\alpha-1}f(t)dt\right].$$

From the Dirichlet formula [4], we have

$$\int_t^a(x-t)^{\beta-1}dx\int_x^a(s-x)^{\alpha-1}f(s)ds = B(\alpha,\beta)\int_t^a(s-t)^{\alpha+\beta-1}f(s)ds,$$

where $B(\alpha,\beta)$ is the beta function. Allowing $a \to \infty$ after using this identity, we have

$$W_x^{\beta}W_x^{\alpha}f(x) = \frac{1}{\Gamma(\alpha)}\frac{1}{\Gamma(\beta)}B(\alpha,\beta)\int_x^{\infty}(t-x)^{\alpha+\beta-1}f(t)dt,$$
$$= \frac{1}{\Gamma(\alpha+\beta)}\int_x^{\infty}(t-x)^{\alpha+\beta-1}f(t)dt,$$
$$= W_x^{\alpha+\beta}f(x).$$

Example 6.4.3 Weyl fractional integral of $f(x) = e^{-ax}$, $a > 0$.

From the definition of Weyl fractional integral, we have

$$W_x^{\alpha}e^{-ax} = \frac{1}{\Gamma(\alpha)}\int_x^{\infty}(t-x)^{\alpha-1}e^{-at}dt,$$
$$= \frac{1}{\Gamma(\alpha)}a^{-\alpha}e^{-ax}\int_0^{\infty}y^{\alpha-1}e^{-y}dy$$
$$\text{(by changing the variable } y = a(t-x)),$$
$$= \frac{1}{\Gamma(\alpha)}a^{-\alpha}e^{-ax}\Gamma(\alpha),$$
$$= a^{-\alpha}e^{-ax}.$$

Example 6.4.4 Using elementary calculus, we can find the Weyl integral of the following functions.
(i) If $f(x) = \cos ax$, $a > 0$, then

$$W_x^\alpha \cos ax = a^{-\alpha} \cos(ax + \frac{\alpha\pi}{2}).$$

(ii) If $f(x) = \sin ax$, $a > 0$, then

$$W_x^\alpha \sin ax = a^{-\alpha} \sin(ax + \frac{\alpha\pi}{2}).$$

(iii) If $f(x) = x^{-\beta}$, $0 < \alpha < \beta$, $x > 0$, then

$$W_x^\alpha x^{-\beta} = \frac{\Gamma(\beta - \alpha)}{\Gamma(\beta)} x^{\beta-\alpha}.$$

Example 6.4.5 Weyl integral for product of x with $f(x)$.

By applying the definition, we have

$$
\begin{aligned}
W_x^\alpha[xf(x)] &= \frac{1}{\Gamma(\alpha)} \int_x^\infty (t - x)^{\alpha-1}[tf(t)]dt \\
&= \frac{1}{\Gamma(\alpha)} \int_x^\infty (t - x)^{\alpha-1}[t - x + x]f(t)dt \\
&= \alpha W_x^{\alpha+1} f(x) + x W_x^\alpha f(x)
\end{aligned}
$$

In particular, if $f(x) = e^{-ax}$, $a > 0$, then

$$W_x^\alpha[xe^{-ax}] = a^{-(\alpha+1)}(\alpha + ax)e^{-ax}.$$

Example 6.4.6 Weyl fractional derivative of $f(x) = e^{-ax}$.

From the definition,

$$D_-^\alpha e^{-ax} = (-1)^n \frac{d^n}{dx^n} W_x^{n-\alpha} e^{-ax},$$

where n is the smallest integer greater than $\alpha > 0$. Let $\beta = n - \alpha$. Then

$$
\begin{aligned}
D_-^\alpha e^{-ax} &= (-1)^n \frac{d^n}{dx^n} W_x^\beta e^{-ax} \\
&= (-1)^n \frac{d^n}{dx^n}[a^{-\beta} e^{-ax}] \\
&= a^{-\beta}[a^n e^{-ax}] \\
&= a^\alpha e^{-ax}.
\end{aligned}
$$

Example 6.4.7 Marchuad fractional derivative of $f(x) = e^{ax}$.

By definition, we have

$$
\begin{aligned}
D_+^\alpha e^{ax} &= \frac{1}{\Gamma(1-\alpha)} \int_0^\infty \frac{e^{ax} - e^{a(x-t)}}{t^{1+\alpha}} dt \\
&= \frac{e^{ax}\alpha}{\Gamma(1-\alpha)} \int_0^\infty \frac{1 - e^{-at}}{t^{1+\alpha}} dt \\
&= \frac{e^{ax}\alpha a^\alpha}{\Gamma(1-\alpha)} \int_0^\infty \frac{1 - e^{-\tau}}{\tau^{1+\alpha}} d\tau \\
&= a^\alpha e^{ax}
\end{aligned}
$$

since, integration by parts gives

$$
\int_0^\infty \frac{1 - e^{-\tau}}{\tau^{1+\alpha}} d\tau = \frac{1}{\alpha} \int_0^\infty \frac{e^{-\tau}}{\tau^\alpha} d\tau = \frac{\Gamma(1-\alpha)}{\alpha}.
$$

As a consequence, e^{ax} is the solution of the fractional differential equation

$$
D_+^\alpha f(x) = a^\alpha f(x).
$$

Example 6.4.8 Liouville and Marchuad fractional derivatives

We know that $D_+^\alpha f(x) = DI_+^{1-\alpha} f(x)$. So

$$
\begin{aligned}
D_+^\alpha f(x) &= \frac{1}{\Gamma(1-\alpha)} \frac{d}{dx} \int_{-\infty}^x (x-t)^{-\alpha} f(t) dt \\
&= \frac{1}{\Gamma(1-\alpha)} \frac{d}{dx} \int_0^\infty t^{-\alpha} f(x-t) dt \\
&= \frac{\alpha}{\Gamma(1-\alpha)} \int_0^\infty \left[f'(x-t) \int_t^\infty \frac{d\eta}{\eta^{1+\alpha}} \right] dt
\end{aligned}
$$

and interchanging the order of integration, we have

$$
D_+^\alpha f(x) = \frac{\alpha}{\Gamma(1-\alpha)} \int_0^\infty \frac{f(x) - f(x-t)}{t^{1+\alpha}} dt, 0 < \alpha < 1.
$$

Here, f' denotes the first derivative of f with respect to its argument. Similarly, we get

$$
D_-^\alpha f(x) = \frac{\alpha}{\Gamma(1-\alpha)} \int_0^\infty \frac{f(x) - f(x+t)}{t^{1+\alpha}} dt, 0 < \alpha < 1.
$$

Example 6.4.9 Generalized fractional derivative (R-L sense) of order $0 < \alpha < 1$
of the function $f(x) = x^\mu$, $\mu \in \mathbb{R}$

From the definition, we have

$$D_{0+}^{\alpha,\rho} x^\mu = \frac{\rho^\alpha}{\Gamma(1-\alpha)} \left(x^{1-\rho} \frac{d}{dx} \right) \int_0^x \frac{s^{\rho-1} s^\mu}{(x^\rho - s^\rho)^\alpha} ds$$

To evaluate the integral, use substitution $t = s^\rho/x^\rho$, we obtain

$$\int_0^x \frac{s^{\rho-1} s^\mu}{(x^\rho - s^\rho)^\alpha} ds = \frac{x^{\mu+\rho(1-\alpha)}}{\rho} \int_0^1 \frac{t^{\frac{\mu}{\rho}}}{(1-t)^\alpha} dt,$$

$$= \frac{x^{\mu+\rho(1-\alpha)}}{\rho} B(1 - \alpha, 1 + \frac{\mu}{\rho})$$

where B is the beta function. Thus, by using the relation between beta and gamma
functions, we obtain

$$D_{0+}^{\alpha,\rho} x^\mu = \frac{\Gamma(1 + \frac{\mu}{\rho})\rho^{\alpha-1}}{\Gamma(1 + \frac{\mu}{\rho} - \alpha)} x^{\mu-\alpha\rho}.$$

When $\rho = 1$, we recover the R-L derivative of x^μ as

$$D_{0+}^{\alpha} x^\mu = \frac{\Gamma(1 + \mu)}{\Gamma(1 + \mu - \alpha)} x^{\mu-\alpha}.$$

6.5 Exercises

6.1. Find the Grunwald–Letnikov derivative of the function $f(x) = (x - a)^2$.

6.2. Find the Hadamard derivative of the function $f(x) = x$ with $a = 1$.

6.3. For the Hadamard fractional integral I_+^α and derivative D_+^α establish that

$$D_+^\alpha I_+^\alpha f(x) = f(x)$$

and

$$I_+^\alpha I_+^\beta f(x) = I_+^{\alpha+\beta} f(x).$$

6.4. Show that the Hadamard operator is a linear operator.

6.5. Identify the other linear operators of fractional derivative and integral.

6.6. Express Hilfer fractional derivative in terms of R-L derivative.

6.7. Express Hilfer fractional derivative in terms of Caputo derivative.

6.8. What is the connection between the Leibniz rule and fractional derivative?

6.9. Prove that the generalized fractional integral operator $^\rho I_{a+}^\alpha$ satisfies the semigroup property.

6.10. If $g'(t) \neq 0$, then prove that

$$I_{a^+,g}^{\alpha} I_{a^+,g}^{\beta} f(x) = I_{a^+,g}^{\alpha+\beta} f(x).$$

6.11. Compute the Hadamard integral of the function $E_{\alpha}[(\ln \frac{t}{a})^{\alpha}]$.

6.12. Find the Weyl fractional derivative of $\cos at$ of order $0 < \alpha < 1$.

References

1. Das, S.: Functional Fractional Calculus for Systems Identifications and Controls. Springer, New York (2008)
2. Diethelm, K.: The Analysis of Fractional Differential Equations. Springer-Verlag, New York (2010)
3. Mainardi, F.: Fractional Calculus and Waves in Linear Viscoelasticity: An Introduction to Mathematical Models. Imperial College Press, London (2010)
4. Miller, K., Ross, B.: An Introduction to the Fractional Calculus and Fractional Differential Equations. John Wiley and Sons Inc, New York (1993)
5. Podlubny, I.: Fractional Differential Equations. Academic Press, San Diego (1999)
6. Sabatier, J., Agrawal, O.P., Tenreiro Machado, J.A.: Advances in Fractional Calculus. Springer, Dordrecht (2007)
7. Yang, X.J.: General Fractional Derivative; Theory. Methods and Applications. CRC Press, Boca-Raton (2019)
8. Kilbas, A.A., Srivastava, H.M., Trujillo, J.J.: Theory and Applications of Fractional Differential Equations. Elsevier, Amsterdam (2006)
9. Samko, S.G., Kilbas, A.A., Marichev, O.I.: Fractional Integrals and Derivatives. Theory and Applications. Gordon and Breach Science Publishers, Amsterdam (1993)
10. Kiryakova, V.: Generalized Fractional Calculus and Applications. Wiley and Sons, New York (1994)
11. Kiryakova, V.: A brief story about the operators of the generalized fractional calculus. Frac. Calc. Appl. Anal. **11**, 203–220 (2008)
12. Katugampola, U.N.: A new approach to generalized fractional derivatives. Bull. Math. Anal. Appl. **6**, 1–15 (2014)
13. Caputo, M.: Linear model of dissipation whose Q is almost frequency independent-II. Geophys. J. Roy. Astron. Soc. **13**, 529–539 (1967)
14. Davison, M., Essex, C.: Fractional differential equations and initial value problems. Math. Sci. **23**, 108–116 (1998)
15. Coimbra, C.F.M.: Mechanics with variable-order differential operators. Annalen der Physik **12**, 692–703 (2003)
16. Trujillo, J.J., Rivero, M., Bonilla, B.: On a Riemann-Liouville generalized Taylor's formula. J. Math. Anal. Appl. **231**, 255–265 (1999)
17. Osler, T.J.: Leibniz rule for fractional derivatives generalized and an application to infinite series. SIAM J. Appl. Math. **18**, 658–674 (1970)
18. Hilfer, R.: Applications of Fractional Calculus in Physics. World Scientific, Singapore (2000)
19. Caputo, M., Fabrizio, M.: A new definition of fractional derivative without singular kernel. Progr. Frac. Differ. Appl. **1**, 73–85 (2015)
20. Atangana, A., Baleanu, D.: New fractional derivatives with non-local and non-singular kernel: Theory and application to heat transfer model. Ther. Sci. **2**, 763–769 (2016)
21. Abdelhakim, A.A.: The flaw in the conformable calculus: it is conformable because it is not fractional. Frac. Calc. Appl. Anal. **22**, 242–254 (2019)

22. Abdelhakim, A.A., Tenreiro Machado, J.A.: A critical analysis of conformable derivative. Nonlinear Dyn. **95**, 3063–3073 (2019)
23. Diethelm, K., Garappa, R., Giusti, A., Stynes, M.: Why fractional derivatives with nonsingular kernels should not be used? Frac. Calc. Appl. Anal. **23**, 610–634 (2020)
24. Giusti, A.: A comment on new definitions of fractional derivative. Nonlinear Dyn. **93**, 1757–1763 (2018)
25. Ortigueira, M.D., Martynyuk, V., Fedula, M., Tenreiro Machado, J.: The failure of fractional calculus operators in two physical systems. Frac. Calc. Appl. Anal. **22**, 255–270 (2019)
26. Ortigueira, M.D., Tenreiro Machado, J.: What is a fractional derivative? J. Comput. Phys. **293**, 4–13 (2015)
27. Sales Teodoro, G., Tenreiro Machado, J.A., Capelas de Oliveira, E.: A review of definitions of fractional derivatives and other operators. J. Math. Phys. **388**, 195–208 (2019)
28. Khalil, R., Horani, M.A., Yousef, A., Sababheh, M.: A new definition of fractional derivative. J. Comput. Appl. Math. **264**, 65–70 (2014)
29. Vanterlal C. Sousa, J., Capelas de Oliveira, E.: On the local M-derivative. Progr. Frac. Differ. Appl. **4**, 479–492 (2018)

Index

© The Editor(s) (if applicable) and The Author(s), under exclusive license
to Springer Nature Singapore Pte Ltd. 2023
K. Balachandran, *An Introduction to Fractional Differential Equations*, Industrial and
Applied Mathematics, https://doi.org/10.1007/978-981-99-6080-4

Printed in the United States
by Baker & Taylor Publisher Services

Printed in the United States
by Baker & Taylor Publisher Services